FIRE ASSAYING

A Practical Treatise on the Fire Assaying of
Gold, Silver and Lead, Including Descrip-
tion of the Appliances Used

BY
EVANS W. BUSKETT, B.S.

Mo. School of Mines, 1895
Chemist, Ozark Smelting and Mining Company
Coffeyville, Kansas

NEW YORK
D. VAN NOSTRAND COMPANY
LONDON
E. & F. N. SPON, Limited
57 HAYMARKET, S. W.
1907

The Plimpton Press Norwood Mass. U.S.A.

PREFACE

THE object of this work is to impart to students the fundamental principles of fire assaying, together with a few practical processes of ore testing. Since the appearance of these articles in "Mines and Minerals" they have been rewritten, and such material added as to bring them up to date.

This book is not intended to supersede any of the more extensive works on this subject, but rather to prepare the student to make accurate assays and ore tests. It will be found especially valuable to superintendents and others connected with mines, who have not had the advantage of a technical education.

The absence of cuts of modern assaying appliances from the larger works on the subject has prompted the author to fully illustrate this book with cuts of the latest improved apparatus. The author wishes to thank the manufacturers of the appliances shown for their kindness in furnishing cuts. Also F. W. Braun & Co., Los Angeles, California, for the use of the tables forming the appendix.

In preparing this book the author has consulted the leading works on assaying, and to this information has added such material from his own experience as he deemed of value.

The author desires to acknowledge his indebtedness to the Helena (Montana) Public Library, also to Mr. B. H. Tatem, assayer in charge of the Helena U. S. Assay Office, and members of his staff, for courtesies extended in allowing him to inspect their office, and for information regarding their methods.

iii

To

My Wife

This book is dedicated in grateful recognition of her ever ready and
inspiring sympathy

CONTENTS

I

SAMPLING

Crushers. Sampling is one of the most important operations in the assay of ores. Samples taken by the engineer. in the field may have been taken very carefully, and if so, represent the actual value of the mine or prospect from which they were taken. They may be "grab" samples, which, as a rule, are not representative.

However, it is the assayer's business to get a fair sample for assay out of the ore as he receives it.

In sampling, cleanliness is imperative. After sampling an ore, the crusher, bucking-board, muller, and all other tools which have been used in the process, should be carefully cleaned before treating the next sample. If a very rich sample is treated and the tools are not properly cleaned, the value of the next sample will be increased. This is called salting.

Samples as they come to the assayer are generally coarse and must be crushed before sampling. They are generally crushed in a small hand-power crusher, Fig. 1, but the crushing may be done in a large iron mortar, Fig. 2, or on a die surrounded on three sides by boards lined with sheet iron, Fig. 3. A heavy

1

box with the end knocked out will answer the purpose of a shield. A bucking-board and muller, Fig. 4, or

Fig. 1.

a Bucks mortar, Fig. 5, are generally used for pulverizing.

Fig. 2.

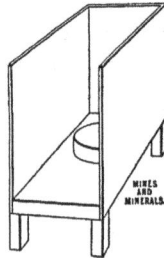

Fig. 3.

Of late, machinery is coming into use for crushing and pulverizing, and most busy assay offices are equipped with these machines, generally driven by electric motors, but often run by hand.

Fig. 4.

Fig. 5.

One of the objections to the older types of crushers was the inaccessibility of the interior crushing surfaces, and consequently they could not be thoroughly cleaned.

This is overcome in the "Lightning" crusher, Fig. 1, manufactured by F. W. Braun and Company,

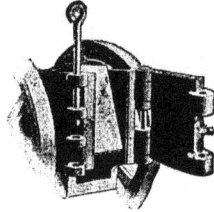

Fig. 6a.

Los Angeles, California. In this crusher the stationary jaw swings on a vertical hinge, exposing both crushing surfaces so that they may be easily and

Fig. 6b.

thoroughly cleaned. These machines will crush fine or coarse and are very efficient when run by power. Fig. 6a shows the machine open, while Fig. 6b is a sectional view.

Figure 7a shows the "Advance" crusher made by The Calkins Company of Los Angeles, California. The front of this crusher raises up for cleaning as shown in the cut, Fig. 7b. This construction is very strong and is free from vibration.

The Case crusher, Fig. 8a, made by the Denver Fire Clay Company, is the latest crusher on the

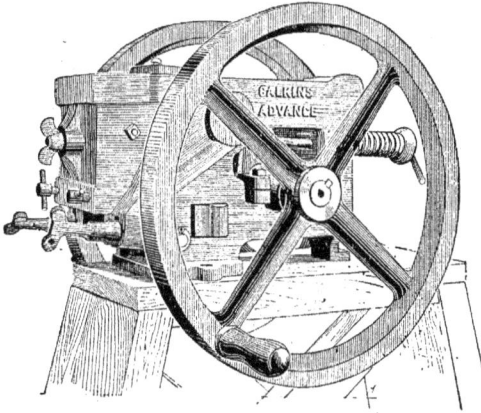

FIG. 7a.

market. It can be operated by hand or power. The frame of this crusher is solid. This construction is very strong and is entirely free from vibration. The adjustment is made by inserting shims between the front jaw and the frame. It is opened for cleaning by turning the hand wheel, lifting out the front jaw and shims, and then raising the movable jaw out of the frame as shown in Fig. 8b. This exposes the crushing surfaces for cleaning and also gives access to the interior of the machine.

Sample Grinders. The Braun disc pulverizer, shown in Figs. 9a and 9b, consists of a stationary disc and a rapidly rotating disc. It is for power

FIG. 7b.

FIG. 8a.

only, very strongly constructed, and all parts are easily accessible for cleaning. It will take a reasonably coarse sample and will reduce one pound of granite to 100 mesh in one minute.

The "Advance" disc sample grinder, Figs. 10a and 10b, made by The Calkins Company, differs from other machines of this type in that it can be driven by hand or power. It also has a gyratory motion in addition to its rotary motion. The rotating disc

Fig. 8b.

is carried on a rocker arm which swings out of the way of the stationary disc, giving full access to the faces of both discs and to all parts of the machine. This machine is strongly constructed, the only wearing parts being the grinding discs, which are easily and cheaply replaced. This machine, operated by hand power, will reduce one pound of coarse granite to 80 mesh in two minutes. Operated by motor power, the capacity is greatly increased.

Hand Samples. Crush the sample until it will pass a ½-inch screen. Shovel the ore into a conical pile and flatten by inserting a flat piece of sheet iron and

Fig. 9a.

Fig. 9b.

moving it in a circle. Divide the circular pile into four parts and take two of the opposite quarters, discarding the other two. Repeat this operation until about one pound of ore is left. Crush the sample to ¼ inch or less, and quarter until about four

FIG. 10a.

FIG. 10b.

ounces of ore are left. Pulverize this on the bucking-board, or in a sample grinder, and pass through an 80-mesh screen. Screens may be had single as shown in Fig. 11, or in sets from 20 to 100 mesh as shown in Fig. 12.

Sampling for Tests. Samples to be tested by amalgamation, chlorination, or cyanide, should not contain less than ten pounds. Crush the sample to $\frac{1}{2}$ inch and quarter it until about five pounds are left. This is passed through a 20-mesh screen, thoroughly mixed,

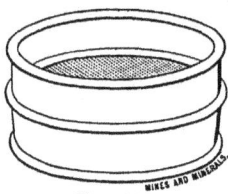
FIG. 11.

and quartered twice. The sample, now weighing about one pound, is put through a 40-mesh screen and a sample "cut out" of it for assaying, which is marked "heads."

Metallic Scales. In pulverizing ores there is sometimes a metallic scale which will not pass through the screen, but which cannot be disregarded, as it is generally very rich. The method for treating metallic scales is illustrated in the following example:

Suppose that the sample weighs four ounces, and the scale, after cupelling and parting, yields 100 milli-

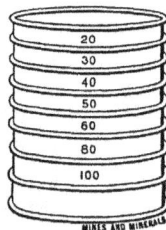
FIG. 12.

grams of gold, then one pound of ore will yield 4×100 or 400 milligrams, and one ton $400 \times 2,000$, or 800,000 milligrams. One ounce Troy = 31,103.5 milligrams; $800,000 \div 31,103.5 = 25.7$ ounces per ton.

Suppose that the ore which passed through the screen assayed five ounces per ton; then one ton of ore will yield 25.7 ounces + 5 ounces, or 30.7 ounces gold.

In assaying lead slags the weight of the metallic scales has to be considered, and the scales are generally assumed to contain 100 per cent lead.

Suppose that the pulp which passed the screen weighed $3\frac{1}{2}$ ounces and contained 70 per cent lead, and the scales weighed $\frac{1}{2}$ ounce. Then $3.5 \times 70 = 2.45$ ounces lead, $.5 \times 1.00 = .50$ ounces lead; therefore, four ounces contain 2.95 ounces lead. $4 : 2.95 :: 100 : x$; $x = 73.75$ per cent lead in the ore.

Making Cupels. Cupels are made from boneash in the following manner: Sift boneash through a flour sifter or a 20-mesh screen. If a one-quart flour sifter is used, three sifters full will make one hundred $1\frac{1}{4}$-inch cupels. Add water, mixing it thoroughly with the ash until the mass will hold together when pressed in the hand. Place the cupel mold, Fig. 13, on the buckingboard, fill with moistened ash, insert the die and strike with two or three sharp blows with a wooden mallet, Fig. 14, slide the mold on the board to prevent sticking, invert and tap the handle on the board to start the cupel. Force the cupel out with a twist of the handle, also turn the cupel when removing it from the die. Place the cupels on a board bottom upward, and allow them to dry in a warm

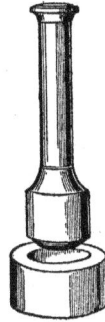

Fig. 13.

room.   Cupels should be dried at least a week before
using.

Cupel Machines.   In well-equipped assay offices,
cupels are not made by hand.   There are three makes

Fig. 14.

of cupel machines on the market.   Fig. 15 shows
Braun's table machine.   This is a very efficient
machine; two hundred cupels an hour can be made
with it.   It is strong built and has a powerful lever-

Fig. 15.

age.   The boneash is placed in the tray and scraped
into the mold with a spatula or with the hand.
The lever is then pressed down and the cupel is
made; it is expelled by moving the bottom gate to

one side and pressing the lever farther down. This machine also is made to bolt to the wall or to a post.

Braun's "automatic" cupel machine, shown in Fig. 16, makes six hundred cupels an hour. This is an expensive machine, and is used only in large works, and by those manufacturing cupels for the market. It has the same lever arrangement as the

Fɪɢ. 16.

smaller machines. The boneash is fed from a hopper to the mold. There are four molds contained in a disc which rotates on the central standard. The molds pass under the hopper and are filled with boneash, the rotation of the disc brings them under the die, and the cupel is made and expelled in the same manner as in the smaller machine.

The "Advance" cupel press, made by The Calkins

Company, Fig. 17, differs from the preceding machine, in that there is but one mold, which passes to and from the hopper. The mold is pushed around under the hopper, carrying with it the bottom gate. These are then returned under the die by means of the handle of the bottom gate. The compression is then made, the bottom gate moved aside, and the

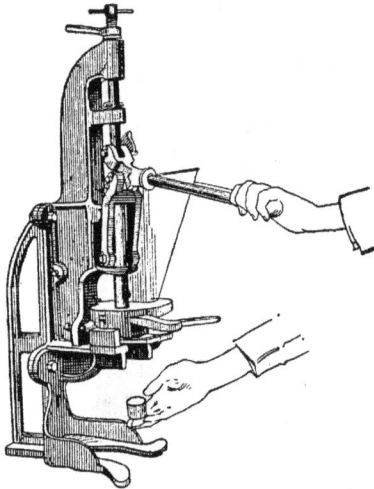

FIG. 17.

cupel expelled. The die has a torsional as well as a vertical movement, which smooths the inside of the cupel and also keeps the die clean.

Ilers' cupel machine, made by the Denver Fire Clay Company, is shown in Fig. 18. This is a foot-power machine, the pressure being applied to a powerful lever by the operator's foot. The die is on the top

of a heavy plunger, and fits into the mold. The
mold is filled with boneash and covered with the top
slide. The foot is then pressed on the lever and the
cupel is made. It is expelled by moving the top
slide to one side and pressing the lever further down.

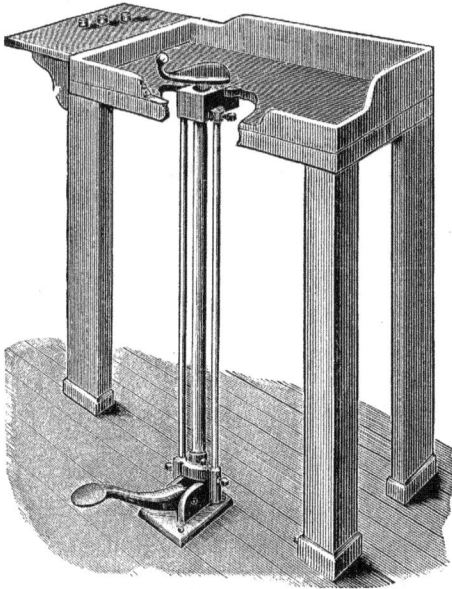

Fig. 18.

The lever is then released and the weight of the
plunger brings the die to its original position; it is
now ready for refilling. This machine has the advantage of allowing the operator to use both hands in
handling the cupels.

II

REAGENTS AND FLUXES

REAGENTS AND REACTIONS

Sodium bicarbonate, $NaHCO_3$, and potassium bicarbonate, $KHCO_3$, are similar in their reactions and may replace each other in fluxes.

Sodium bicarbonate is a basic flux, a desulphurizing agent, and in some cases an oxidizing agent. It is a powerful solvent and is capable of carrying a large amount of material in suspension and remaining fluid. It absorbs lime, alumina, and many other oxides.

When sodium bicarbonate is heated, it loses water and carbon dioxide and becomes sodium carbonate.

$$2NaHCO_3 + heat = Na_2CO_3 + H_2O + CO_2.$$

The liberated carbon dioxide has an oxidizing action on sulphur and the metals.

Sodium carbonate, Na_2CO_3, reacts similarly to sodium bicarbonate, but cannot be recommended for fluxing as it absorbs moisture from the air, and becomes hard and lumpy. Fused with silica it forms sodium silicate and carbon dioxide is given off.

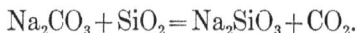

$$Na_2CO_3 + SiO_2 = Na_2SiO_3 + CO_2.$$

16

The escape of the carbon dioxide causes the charge to boil violently.

Potassium carbonate, K_2CO_3, is similar in its reactions to sodium carbonate and with sodium bicarbonate forms the base of nearly all fluxes, as such a mixture forms a double carbonate which fuses at a lower temperature than either alone, and when fused is very liquid and a powerful solvent.

Metallic Lead, Pb, in granulated form, is used in the scorification assay as a collector of gold and silver. Sheet lead is used in cupelling and in the assay of bullion.

Granulated lead may be made as follows: Pour about five pounds of melted lead into a box with handles on the sides, Fig. 19, which should be placed

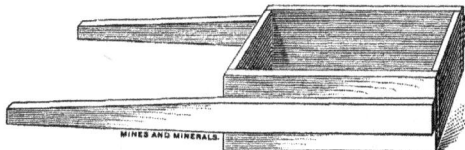

FIG. 19.

on a bucking-board or suspended by wires. Shake slowly until the lead becomes pasty, then fast until it solidifies. The product should be screened through a 20-mesh screen and the coarse material remelted with the next charge.

Test lead should be scorified and cupelled to determine its silver value, which must be deducted from all assays made with it. The amount taken for this test should be the same as that used in the assay of ores, or a multiple of that amount.

Litharge, PbO, is a basic flux and furnishes the lead which collects the gold and silver values from the ores into a button. It is a solvent and also oxidizes sulphur and other easily oxidizable substances. Litharge generally contains an appreciable amount of silver and should be assayed to determine whether or not it can be disregarded. Pueblo litharge is practically free from silver.

The presence of red lead is detrimental to litharge as it is said to oxidize silver. Litharge can be freed from red lead by fusing it and allowing it to cool in a closed vessel.

Fused with silica, litharge forms a lead silicate.

$$PbO + SiO_2 = PbSiO_3.$$

About three and one-half parts of litharge to one of silica forms lead silicate, which is easily fusible and fluid when fused. Metallic oxides are dissolved by litharge and unite with silica of the crucible if there is not enough silica in the charge to neutralize them.

Borax, $Na_2B_4O_7$, $10H_2O$, is an acid flux containing nearly 50 per cent of water. It is used in lead fluxes and also as a cover.

Borax glass is made by fusing borax in a large crucible and pouring the fused material on a clean surface of iron, brick, or other fireproof material. It is used in the scorification assay as a solvent for metallic oxides.

Borax glass has been recommended for use in fluxes, in preference to borax which swells when heated, but it is more violent in its action than borax, causing much boiling. It also absorbs moisture from the air, making the flux hard and lumpy.

Silica, SiO_2, is an acid flux and is used with ores containing lime, aluminum, iron, etc., forming fusible silicates. Ground window-glass can be used as a substitute for silica.

Iron, Fe, in the form of nails or wire, is used in the assay of lead, and is sometimes used in the gold and silver assay of sulphide ores, but is not to be recommended.

Nails with silica can be used to advantage with ores containing lime, magnesia, etc., as the iron forms double silicates with these substances which are more easily fusible than their single silicates.

Wheat flour is a strong reducing agent, one gram of flour reducing about fifteen grams of lead from litharge. It is quiet in its action, causing no ebullition.

Argol, $KHC_4H_4O_6$, an impure potassium bitartrate, is a reducing agent and a basic flux. When heated it produces potassium carbonate and carbon. One gram of argol will reduce from five to eight grams of lead from litharge.

Charcoal is sometimes used as a reducing agent, one gram of charcoal reducing from twenty to thirty grams of lead from litharge.

To find the reducing power of any reducing agent the following charge should be used:

Litharge ....30 grams
Soda, or lead flux 20   "
Reducing agent............ 1   "

Melt, pour, and weigh the resulting lead button, the weight of which will be the reducing power.

Potassium cyanide, KCN, is a powerful reducing and desulphurizing agent. It has great affinity for oxygen forming cyanate.

$$KCN + PbO = KCNO + Pb.$$

It combines with sulphur forming the sulphocyanate.

$$KCN + PbS = KCNS + Pb.$$

Potassium cyanide is very poisonous and should be handled with great care.

Potassium nitrate, KNO_3, is a powerful oxidizing agent and basic flux and is used in the assay of sulphide, telluride, and arsenide ores. It melts at about 350 degrees and decomposes at a higher temperature, yielding oxygen, which oxidizes sulphur and many of the metals. One gram of nitre will oxidize about four grams of lead. The oxidizing power is determined with the following charge:

Litharge ....30 grams
Soda, or lead flux 20   "
Flour..................... 1   "
Nitre..................... 2   "

Melt, pour, and weigh the resulting lead button. Suppose the reducing power of the flour is 15, and the lead button weighs 7 grams, then $15 - 7 = 8$ grams lead oxidized by two grams of nitre, or one gram will oxidize four grams of lead.

Salt, NaCl, is used as a cover in the assay of lead and also with certain classes of gold and silver ores. Heavy iron sulphides and earthy ores will form a more fluid slag when covered with salt than with borax cover. Salt melts at a low temperature, forming a liquid cover which prevents ebullition and also oxidation from the air. A teaspoonful of salt added to a boiling assay will cool the charge and prevent boiling over.

Slags

Assaying is smelting on a small scale. Therefore the reactions which take place in the crucible are almost identical with those in the smelting of ores in the blast furnace, and the rules for fluxing ores in the furnace may be applied to the fluxing of ores in the crucible. However, the proportion of flux to ore is much greater than in smelting and consequently the fluxing does not require as close calculation, and the amounts of the different fluxes are only approximated, the excess of soda and potassium carbonate generally balancing any disproportion.

In the assay of gold and silver ores the lead is reduced from the litharge by the reducing agent combining with the oxygen of the litharge,

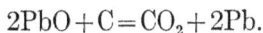

$$2PbO + C = CO_2 + 2Pb.$$

This metallic lead coming in contact with the ore at the moment of its reduction, and during its subsequent boiling, forms an alloy with the gold and silver which sinks to the bottom of the crucible as soon as the boiling stops. This boiling is caused by the carbon dioxide escaping from the charge.

It is the object of the assayer to bring the lead in contact with every particle of gold and silver in the charge, and to fulfil these conditions the slag should not be too thick nor too fluid, nor should it melt too quickly. The best results are obtained with a flux which will bring the lead into the most intimate contact with the ore for a reasonable length of time, and at the same time produce a slag which is fluid enough to pour cleanly.

In order to do this the reactions of the different substances which make up the flux should be known and also the composition of the ores.

Acid ores are ores containing a large amount of uncombined silica. These ores are easily fluxed with soda and a large amount of litharge, the soda and part of the litharge forming fluid silicates,

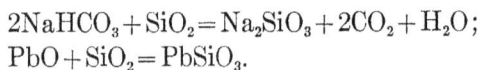

$$2NaHCO_3 + SiO_2 = Na_2SiO_3 + 2CO_2 + H_2O;$$
$$PbO + SiO_2 = PbSiO_3.$$

Large quantities of water and carbon dioxide are given off, and unless an excess of litharge is used the crucible will probably boil over.

These two reactions, that is, the reduction of lead from litharge and the formation of silicates, are the principal reactions which take place in the assay of ores, basic or acid. Ores containing sulphur, tellurium, etc., have other reactions and require a different treatment. This matter will be discussed in the chapter on the assay of base ores.

FLUXES

Western assayers generally use a lead flux as a base, adding litharge and flour, and other reagents, as the ore may require. This will be found advantageous in custom offices where a great variety of ores are assayed. Where the same ore is being assayed daily, it is more convenient to make up the flux completely, adding the litharge and flour in bulk.

Lead flux:

> Sodium bicarbonate . . . . 15 parts
> Potassium carbonate . . . . 10 "
> Powdered borax . . . . 5 "

This may be made the base of a flux for lead and also gold and silver ores, adding flour for lead ores, litharge and flour for acid and earthy ores, and litharge with the proper amount of nitre for sulphides, etc.

General formulas are sometimes used, adding nails when the ore contains sulphur. The addition of nails is not to be recommended with heavy sulphides, as the button almost invariably must be scorified before cupelling.

Aaron's general formula:

Ore 1   assay ton
Litharge ....1½   "   "
Soda 3   "   "
Borax ....½   "   "
Flour ....½   "   "
Iron ....1 to 3 nails
Salt ....Cover

The following table of charges is made up on the basis of the lead flux given above.

Ore	Character of the Gangue	Assay, Tons Ore	Grams of Lead Flux No. 1	Grams of Soda	Grams of Litharge	Grams of Silica	Grams of Flour	Iron Nails	Cover
Oxidized	Neutral, No. *Pb*	½	20		60		1.5		Borax
Oxidized	Lime, alumina, baryta	½	20		30	15	1.5	2	Salt
Oxidized	Iron	½	20		30	10	1.5		Borax
*Galena	Lead, 80%	½	20		30				Borax
*Galena	Siliceous lead, 30%	½	20		15				Borax
Lead Carbonate	Neutral lead, 40%	½	20		15		1.5		Borax
Quartz	No bases	½	20	10	60		1.5		Borax
*Iron pyrites	Concentrates	½	20		90	15			Salt
*Lead matte		½	20	15	20	15			Salt
*Copper matte		½	20	15	30	15			Salt
*Tellurides	Siliceous	½	20		60				Borax
*Arsenical	Iron Sulphides	½	20		40				Borax
Tailings		1	40		30		1.5		Borax
Slags		1	40		30		1.5		Borax

If buttons from sulphide assays are hard and brittle, or if any matte is formed, the buttons, together with the matte, should be scorified with test lead and borax glass.

Lead buttons should not weigh more than twenty grams, as this is a convenient size for cupelling and

*The quantity of nitre should be determined by preliminary assay.

is large enough to insure a complete extraction of values. Large buttons require a longer cupellation, resulting in a consequent loss of values by volatilization.

III

ASSAY OF ACID ORES

Assay Ton System. In the United States ore is weighed in tons of 2,000 pounds, while bullion is weighed in troy ounces. Hence we have the assay ton system, which simplifies the work of the assayer, freeing him from the necessity of calculating each assay separately.

The system is based on the proportionate weight of one ounce troy to one ton avoirdupois:

1 pound = 7,000 grains troy.

2,000 pounds = 14,000,000 grains troy.

480 grains troy = 1 ounce troy.

$$\frac{14,000,000}{480} = 29,166 \text{ ounces troy in one ton avoirdupois.}$$

By expressing the number of troy ounces in a ton in milligrams we have a weight containing 29,166 milligrams, which is called the assay ton. Therefore, if we take one assay ton of ore for assay, every milligram of metal extracted from that quantity will represent one ounce troy per ton of 2,000 pounds.

Furnaces. Coal furnaces fired from below are generally used in large assay offices, the usual design being shown in Fig. 20. This furnace is manu-

factured by the Denver Fire Clay Company and
consists of two 9 × 15 muffles, placed one above the
other and surrounded by brickwork. It is called a
two-story furnace. These furnaces are also built
with one muffle about twenty inches square, the opening in front being reduced to about 8 × 5 inches.

Figure 22 shows a portable furnace for coal or coke,

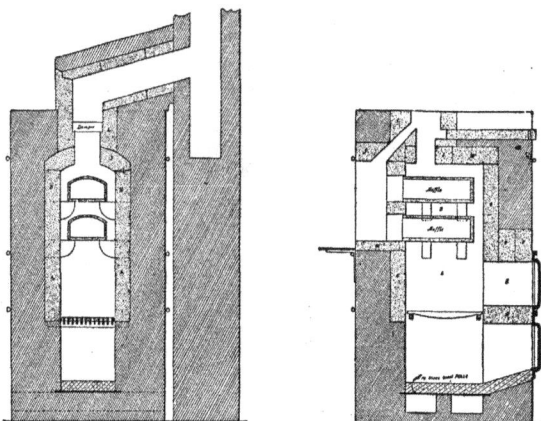

Fig. 20.

made of fire clay banded with iron. These furnaces
were once very popular, but are now superseded by
the gasolene furnace.

Figure 22 shows a 9 × 15 gasolene furnace, and
Fig. 23 the tank, and Figs. 27 and 28 two styles of
burners. These furnaces can be heated very quickly,
thirty or forty minutes sufficing to heat the muffle
to redness.

Figure 24 shows the "Advance" combination fur-

Fig. 21.

Fig. 22.

nace in which the melting is done in a chamber underneath the muffle. This form has the advantage of allowing the assayer to melt one "batch" while he is cupelling another. The cupelling is done with the door closed, as the air is admitted underneath the door and passes through the muffle and into the

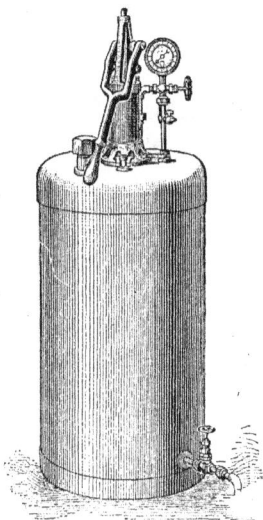

Fig. 23.

stack by means of a special flue shown in Fig. 25. It is manufactured by The Calkins Company of Los Angeles, California.

Figure 26 is Braun's combination furnace, manufactured by F. W. Braun Company, Los Angeles, California. This furnace has the melting chamber at one end and the muffle at the other. Melting and

FIG. 24.

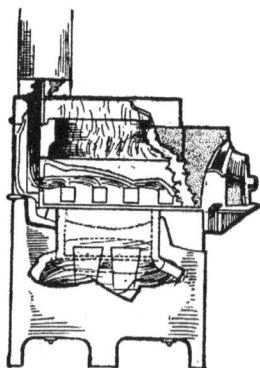

FIG. 25.

cupellation can be carried on at the same time in
this furnace. It can be turned on its central pivot
and the burner presented at either end.

The gasolene burners shown in Figs. 27 and 28
are the general forms of these burners. The gasolene
entering the supply tube passes around the front of
the burner where it is vaporized by the heat, and

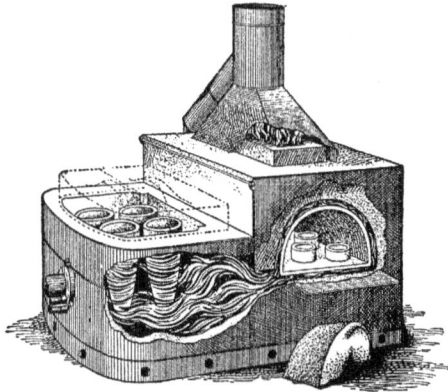

Fig. 26.

returns to the rear as a gas, and is injected into the
furnace through a large tube. Auxiliary heating
tubes are used to heat the gasolene when starting the
burner.

Scorification Assay. This method of assay is par-
ticularly adapted to the assay of rich silver ores, but
cannot be recommended for gold, as the amount of
ore taken is too small for accurate work.

Scorification is an oxidizing operation, the lead
being oxidized by the air and the gangue dissolved

by the borax glass. Some of the litharge formed is taken up by the borax, making the slag more fluid.

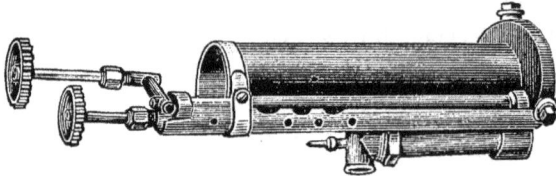

FIG. 27.

Weighing. Pour the sample out on a rubber cloth about ten inches square, and mix thoroughly by raising alternately the corners of the cloth, smooth

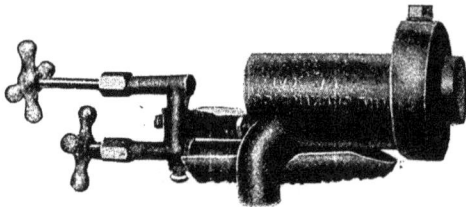

FIG. 28.

out with a spatula, Fig. 29, and weigh out one-tenth assay ton on a pulp balance, Fig. 30. Place the ore in a scorifier, Fig. 31, with 15 grams test lead; mix

FIG. 29.

thoroughly with a spatula and cover with 15 grams test lead and from .5 to 2 grams borax glass. The test lead and borax may be measured.

Place in the muffle with the scorifier tongs, Fig. 32,

and close the door until the charge is thoroughly melted, which is shown by the mirror-like appearance of the metal, surrounded by a ring of borax.

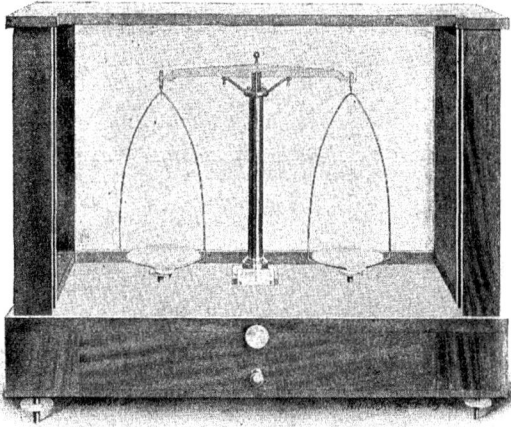

FIG. 30.

The metal within the ring of slag is called the "bull's-eye." Open the door until the ring of slag closes over the "bull's-eye"; when the operation is completed, close the door for five minutes to render the slag more fluid. Remove from the muffle and pour into conical molds. A twelve-hole mold, Fig. 33, is best, as a 9 × 15 muffle (the size generally used) will hold twelve scorifiers.

FIG. 31.

FIG. 32.

When cool, slag the buttons on an anvil, Fig. 34, using a flat-face hammer, Fig. 35, weighing two or three pounds. Hammer the buttons into cubes. They are now ready for cupellation, which will be described under Crucible Assay. If the button is

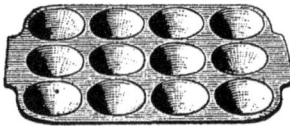

FIG. 33.

too large for cupellation, it should be scorified again with borax glass.

A pinch of flour added when the "bull's-eye" is closed will reduce a part of the lead in the slag, carrying down any suspended values. Sulphide ores

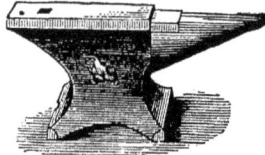

FIG. 34.

should be scorified without borax glass until the "bull's-eye" is closed, then add borax glass with a spoon or in a piece of tissue paper. Litharge is sometimes used as a cover for sulphide ores.

Crucible Assay. Mix the ore as described in the scorification assay and weigh out one-half assay ton, taking dips with the spatula from all parts of the ore.

Place in a ten-gram crucible, Fig. 36, with the following charge:

Lead flux ....20 grams
Litharge ....60 "
Flour ....1½ "

Mix well in the crucible with a spatula, holding

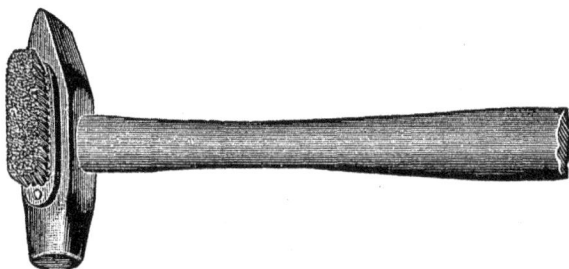

Fig. 35.

the crucible in the left hand and the spatula in the right, turning the crucible while mixing. Cover with a spoonful of powdered borax. Place the crucible in the muffle with the scorifier tongs.

The muffle should be red hot when the crucibles are put into it. The fusion will take from twenty to thirty minutes and the crucibles should be allowed to remain in the muffle about fifteen minutes after fusion, during which time the temperature should be raised, in order to render the slag fluid. If the

Fig. 36.

crucible threatens to boil over add a teaspoonful of salt. (A saltspoon can be made by binding together

with wire the handle of a tin spoon and a round bar
of iron about twenty inches long.) Pour the fused
charge into the mold, beginning slowly and ending
quickly. Always tap the crucible on the mold and
turn completely over, so that the drop of slag will
fall inside.

Cupellation. As soon as the crucibles are removed
from the muffle, the cupels, Fig. 37, are put in with

the cupel tongs, Fig. 38, so that they
will be hot when they receive the lead
button. Slag the button and place in
the cupel, which should be red hot, with
the cupel tongs. Close the door of the
muffle until the cupellation begins,

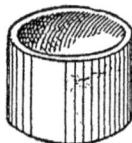

Fig. 37.

which is shown by the fumes of lead oxide that are
given off. The door is now opened to increase the
oxidation by allowing more air to enter. When the
cupellation is nearly finished, bright colors play over
the surface of the button; it will suddenly darken;
then brighten again with a quick flash. It should

Fig. 38.

then be drawn to the front of the muffle and allowed
to cool slowly to prevent "spitting" or "sprouting,"
which is caused by the sudden efflux of oxygen which
was absorbed by the button when hot. When the
buttons are large they should be covered with a hot
cupel, that they may cool very slowly. Results are
not reliable when the button has "spit."

If the temperature is too low, or the draft through the muffle too strong, the buttons will "freeze," *i.e.*, become covered with a film of lead oxide which prevents further oxidation. They can be thawed by placing them in a hotter part of the muffle, or by placing a piece of charcoal on top of the frozen cupel. Results from thawed buttons are not reliable. They are especially unreliable when the cupellation has proceeded for some time before freezing; and when the original lead button is small.

Scorifiers placed in front of the cupels will prevent freezing and will "feather" them nicely. "Feathers" is the assayer's term for the oxide of lead which crystallizes around the inside of the cupel. When a cupel is feathered the temperature has not been high enough to volatilize any of the silver.

Weighing. As the cupels are removed from the

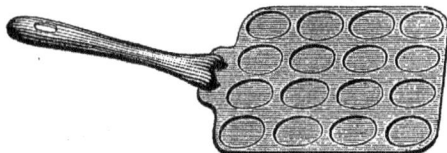

Fig. 39.

muffle they are placed in order on the cupel tray, Fig. 39, and are then ready for weighing.

The weighing is done by very fine balances. Fig. 40 shows one of the latest improved balances manufactured by F. W. Thompson, Denver, Colorado. This balance is sensitive to $\frac{1}{500}$ of a milligram and has powerful magnifying glasses for both beam and

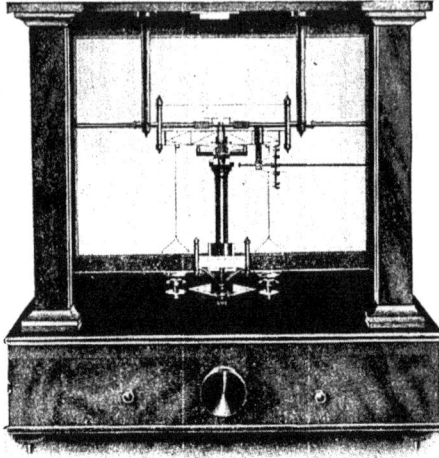

Fig. 40.

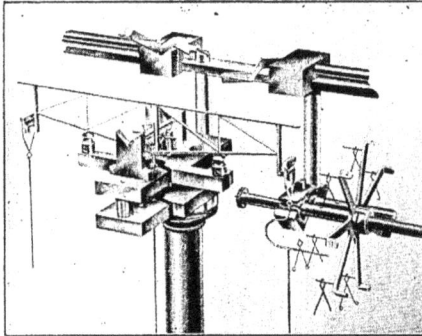

Fig. 41.

needle.   Fig. 41 shows the multiple rider attachment
for weighing without weights.   This device, invented
by Mr. Thompson, consists of riders of different

weights carried on a star wheel. They are distinguished from each other by their peculiar bend, each rider having its characteristic bend, which soon becomes familiar to the operator. The cut shows the construction and operation of the device very clearly. It can be attached to any balance.

Figure 42 shows the Keller portable balance made by the Saltlake Hardware Company. This balance is very light, increasing its sensitiveness, and has a side pointer which permits a small case. Sensitive to $\frac{1}{100}$ of a milligram.

Figure 43 shows a balance of the same make with a top pointer. Sensitive to $\frac{1}{300}$ of a milligram.

A short beam and lightness are the chief requisites of a quick, accurate balance.

The buttons are removed from the cupels with a pair of button pliers, Fig. 44, cleaned with a button brush, Fig. 45, and placed in a row on the base of the balance. They are then weighed in rotation, the weight of each button being recorded as it is weighed. As the buttons are weighed they are put into porcelain crucibles, Fig. 46, for parting. A tray, Fig. 47, made of heavy sheet iron, is very convenient for parting, as all the crucibles can be handled together.

Dilute nitric acid is generally used for parting, although sulphuric acid is sometimes used. Parting acid is usually made of one third chemically pure acid and two thirds distilled water. The acid should be tested for chlorine and rejected if any is found.

Fill the crucibles containing the buttons with acid

FIG. 42.

Fig. 43.

and place the rack on a hot plate. Hot plates may
be heated by gas or gasolene, or an oil-stove, Fig. 48,
may be used.

While the buttons are parting, bubbles of gas are
given off; as soon as they cease the parting is done.
The acid is then poured off and the residue washed
several times with distilled water, by means of a

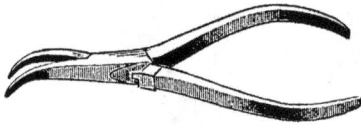

Fig. 44.   Fig. 45.

wash bottle, Fig. 49. Place the edge of the crucible
against the nozzle of the wash bottle and slowly tilt
it until all the acid is poured off. Fill the crucible
with water and repeat the operation. Wash several

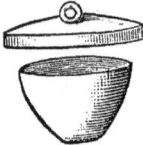
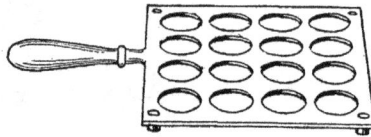

Fig. 46.   Fig. 47.

times and remove the final drop of water with a
small strip of blotting paper. As the assays are
washed they are placed in their respective places in
the rack and dried. When all are thoroughly dry
the heat is raised and the gold brightened.

Buttons, to part easily, should contain at least
twice as much silver as gold. Small buttons, after

weighing, should have more silver added to them.
This may be done by replacing them in the cupels

FIG. 48.

with a small piece of chemically pure silver foil and
melting them together with a blowpipe and alcohol

FIG. 49.   FIG. 50.

lamp, Figs. 50 and 51. This may also be done on
charcoal.

In weighing gold use a pair of jeweler's tweezers, Fig. 52. Loosen the gold with the point of the tweezers and tap the crucible to bring all the gold together. Take off the balance pan; hold the crucible so that the lower edge is just above the pan, and

FIG. 51.

tilt it gradually, tapping the bottom to start the gold. When all the gold has been transferred to the pan, replace it on the balance and weigh. The weight of the gold subtracted from the weight of the button before parting will be the weight of the silver. In handling the balance pan use the tweezers.

FIG. 52.

No part of the balance should be touched with the hands, except when taking it down or setting it up. After being set up, the balance should be brushed with a camel's-hair brush and allowed to stand several hours before using.

IV

ASSAY OF BASE ORES

Base ores may be divided into two general classes: oxide and sulphide ores. Oxidized ores contain lime, magnesia, iron, as oxides or sulphates, and generally some silica.

Sulphide ores contain, besides sulphur, antimony, arsenic, tellurium, etc. Such ores must be fluxed with a view to keep these metals out of the lead button, to produce a button of the proper size for cupellation when one-half assay ton is taken, and at the same time to form a fluid slag. Slags should always pour cleanly, and if there are any shots of lead left in the crucible after pouring the results are not reliable.

Some assayers use nails for desulphurizing, but the general practice in the West is to use nitre. When nails are used the button is almost always too large for the cupel and must be scorified before cupelling. This is a waste of time and fuel, as the assay can be made with nitre and the proper size button reduced.

Nails and silica are used in the assay of sulphates, as baryta, gypsum, etc. These ores can also be assayed with nitre.

Preliminary Assay. Preliminary assays are made

45

to determine the amount of nitre necessary to oxidize
the sulphur in the ore, leaving enough, however, to
reduce a lead button of proper size for cupelling.
For preliminary assays use the following charge:

Ore .... $\frac{1}{10}$ assay ton
Litharge ....30 grams
Lead flux 20   "
Borax ....Cover

Place in a five-gram crucible, melt and pour. In
the regular assay the button should weigh 20 grams.
If one-half assay ton is taken for the regular assay,
and the button from the preliminary weighs 11 grams,
we have $11 \times 5 = 55$ grams, which one-half assay
ton will reduce: $55 - 20 = 35$ grams of lead remaining to be oxidized. Then if the oxidizing power of
the nitre is 4, we have $35 \div 4 = 8.75$, or 8.8, the
number of grams of nitre necessary to reduce a
20-gram button from one-half assay ton of the ore.
In weighing nitre the nearest gram weight is taken,
i.e., if the preliminary assay calls for 8.8 grams,
9 grams are taken.

The following table gives the amounts of nitre to
be used for buttons weighing from 5 to 30 grams.
The oxidizing power of the nitre in this table is 4.

Grams Lead	Grams Nitre	Grams Lead	Grams Nitre
5	1.2	18	17.5
6	2.5	19	18.8
7	3.8	20	20.0
8	5.0	21	21.2
9	6.2	22	22.5
10	7.5	23	23.8
11	8.8	24	25.0
12	10.0	25	26.2
13	11.0	26	27.5
14	12.5	27	28.8
15	13.8	28	30.0
16	15.0	29	31.2
17	16.2	30	32.5

Melting. Weigh one-half assay ton into a ten-gram crucible containing the following charge.

Lead flux ....20 grams
Litharge ....30 "
Nitre ....9 "
Borax ....Cover

If the ore contains much iron, five to ten grams silica should be added. Melt and pour as described under acid ores. If the button is brittle it should be scorified with borax glass. The button may be covered with matte; if so, scorify the button and matte together until the "bull's-eye" is nearly closed, then add borax glass and complete the scorification. If the button is clean and soft, but too large for the cupel, it may be cut into and cupelled in two cupels and the resulting silver buttons weighed and parted together. After a little practice one becomes familiar with an ore and can reduce the proper size button without fail

Ores containing barytes and other sulphates should

be fluxed with silica and iron, using the following
charge:

```
Ore ..................½ assay ton
Lead flux ..............20 grams
Litharge .............30   "
Silica................15   "
Flour ...............1½   "
Nails.................. 2   "
Borax ..............Cover
```

Roasting. Ores containing a large amount of sulphur are sometimes roasted. The roasting should
be done in a clean scorifier. One-half assay ton of
ore is taken and mixed with about five grams of charcoal; the mixture is placed in a scorifier and gradually
heated in the muffle. If heated too quickly there
will be a loss by decrepitation. The ore should be
roasted dead and all the charcoal burned out. Cool
and remove the ore to an agate mortar and break
up the lumps. Brush into a ten-gram crucible
containing the following charge:

```
Lead flux .................20 grams
Litharge ..................60   "
Flour ....................1½   "
Silica....................10   "
Nails....................2
Borax ...................Cover
```

Heavy iron sulphides should be run with a heavy
litharge charge, as follows:

Ore ....½ assay ton
Litharge......90 grams
Silica.........10 "
Nitre.........Amount determined by pre-
                liminary assay
Salt...........Cover

This charge should be run in a fifteen-gram crucible.

Wet Methods. Iron pyrites may be dissolved in nitric acid and the residue scorified.

Weigh out one assay ton of ore and dissolve in nitric acid, using a large casserole. When the solution is complete, pour into a large beaker and dilute to about 1,000 cubic centimeters and add about one gram of salt to precipitate the silver. When the precipitated silver has settled, decant the supernatant liquid and wash the precipitated silver chloride into a filter. Wash thoroughly with hot water, dry, and burn the filter and precipitate in a scorifier. Add thirty grams test lead, one gram borax, and scorify. Cupel and part as in an ordinary assay. Results obtained by this method are generally higher than those obtained in the ordinary crucible or scorifier assay.

Whitehead's Method. Weigh out one assay ton of ore into a large casserole and add gradually enough nitric acid to dissolve it. Boil, to expel red fumes; dilute to 500 cubic centimetres and add fifty grams lead acetate. Stir and add one cubic centimetre of dilute sulphuric acid; allow the precipitated lead sulphate to settle and filter. Burn the filter and

scorify the residue with test lead. Cupel the button,
weigh the gold and silver button, and part. The
filtrate is diluted to 1,000 cubic centimetres; divide
into two parts and precipitate the silver with sodium
bromide with constant stirring. Filter, wash with
cold water, and dry the filter. The precipitate is
brushed from the filter, mixed with about three times
its weight of sodium carbonate and a little flour,
placed in a crucible, covered with borax and melted.
The silver obtained is added to that obtained in the
gold assay. Duplicate assays should agree to within
two tenths of an ounce silver.

Reactions. The reactions of base ores are similar
to those of acid ores. In acid ores the base is sup-
plied by the soda and litharge, while in the case of
basic ores the addition of silica is generally necessary.
Silicates are classified according to the proportions
existing between the amounts of oxygen contained
in the base and acid.

	O In Base	O In Acid
Subsilicate.........................	2	1
Monosilicate.......................	1	1
Disilicate..........................	1	2
Trisilicate	1	3
Sesquisilicate......................	2	3

The subsilicate and monosilicate are the most
easily fusible, while the sesquisilicate is difficultly
fusible.

In assaying it is not necessary to calculate the

silica very closely, as the excess of soda and litharge will take up any reasonable excess of silica. The following table shows the amounts of silica necessary to convert one assay ton of the several bases into slag:

Monosilicates	Silica	Bisilicates	Silica
Lime................	0.535 A T	Lime................	1.070 A T
Magnesia.............	0.750 A T	Magnesia.............	1.000 A T
Alumina.............	0.873 A T	Alumina.............	1.747 A T
Ferrous Oxide.........	0.416 A T	Ferrous Oxide........	0.833 A T
Manganous Oxide......	0.422 A T	Manganous Oxide...	0.845 A T

Magnesia and alumina require much more silica than lime or iron. Clay ores will be made more easily fusible by adding a little iron. Base ores containing sulphur have different reactions, the sulphur forming a matte which is dissolved in the slag.

$$Fe + S = FeS.$$

Iron and copper ores containing sulphur form mattes which are oxidized by the nitre and dissolved in the slag. Iron added to lead assays produces a matte which is dissolved in the slag, the silica in the ore aiding the solution.

V

LEAD ASSAY

The fire assay of lead is the most difficult of fire assays, and it requires close attention and long experience on the part of the assayer in order to produce reliable results. Lead is very easily oxidized and care must be exercised in both the fluxing and melting of lead ores.

Weighing and Fluxing. The flux must contain enough reducing agent to prevent any oxidation and the charge should be easily fusible. The following charge works well on all classes of ores:

Ore......................10 grams
Lead flux ....30 "
Flour .... 2 "
Nails.................... 2
Salt.....................Cover

Melting. The muffle should be at a low, red heat when the crucibles are put in, and allowed to remain at that temperature until the fusion is complete. This generally takes about twenty-five minutes. The heat is then raised for about ten minutes, when the assays are poured. In pouring, the nails are removed with a small pair of tongs, Fig. 53, and are washed in the slag to remove any adhering particles of lead. The buttons are slagged and hammered into cubes, as in the gold and silver assay, and weighed; every 100 milligrams is equal to one per cent lead.

Crucibles may be used several times, but crucibles that have been used for gold and silver assays cannot be used for lead assays as the slag remaining in them contains lead. It is well to use different grades or makes of crucibles for different classes of assays, in order to readily distinguish between them.

Sulphur is added when the ore contains copper oxide or carbonate. The sulphur and copper form

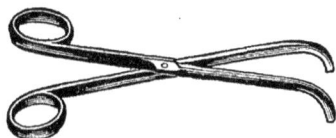

Fig. 53.

a matte which is dissolved by the slag. This prevents the copper from combining with the lead, which would produce a false result.

Wet Method. Copper ores are sometimes dissolved in acid and the residue assayed.

Weigh out ten grams of ore, place in a casserole with nitric acid, and boil until the red fumes are all expelled. Cool and add sulphuric acid, and boil until white fumes are copiously evolved. Cool, dilute, and filter. Wash and dry the precipitate and filter, invert over a clean scorifier and ignite. Brush the residue and carbonized filter into a crucible containing the lead charge, and assay as usual.

Fire assays of lead ores containing antimony are not always accurate, as part of that metal is reduced and alloys with the lead, making the assay run high. Antimony sulphide ores fluxed with a large amount of soda will give a clean lead button.

Arsenic produces a speiss which can generally be easily separated from the lead. Sometimes the lead and speiss button are taken together and assayed for lead by wet methods.

In assaying slag from the lead furnace it is well to add one gram of silver to each assay. The silver acts as a collector of lead.

VI

BULLION ASSAY

Assay of Base Bullion. Base, or lead bullion, is the term applied to the lead product of the lead smelting of gold and silver ores.

Sampling. In the lead smelting of gold and silver ores, the lead, enriched by the gold and silver extracted from the ores, is drawn from the furnace and cast into bars or "pigs." These "pigs" are sampled with a heavy punch which cuts out a piece of bullion about two inches long and one-eighth inch in diameter.

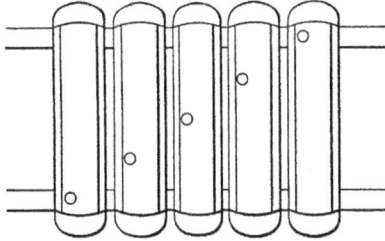

FIG. 54.

The "pigs" are placed in rows of five and a hole punched in each, the holes running diagonally across the bars. They are then turned over and another row of holes are punched diagonally opposite the first row. Fig. 54. As the holes are punched the cuts are placed in a bucket or can. When a lot has

55

been sampled the sample is melted in a clay or graphite crucible. Care must be taken in melting the sample. The heat must not be high enough to volatilize the lead as this would give high results. As soon as the sample is melted it is stirred with an iron rod and poured into a mold about twelve inches long, four inches wide, and one-half inch deep. Four samples are cut from this bar, each weighing one-half assay ton. Fig. 55.

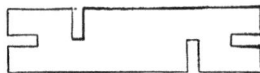

FIG. 55.

Cupelling. The cupels are placed in the muffle and heated before the assays are charged. They should be examined to be sure they are free from cracks. The samples are hammered into cubes and cupelled in the usual manner. The cupels must show feathers, and should be moved to the back of the muffle just before the brightening of the buttons in order to expel the last of the lead.

Parting. After weighing, the buttons are flattened and parted in porcelain crucibles, or, if large, in matrasses or parting flask. Two buttons are parted together. Ordinary parting acid is used and the solution is boiled until all nitrous fumes are expelled. The acid is then diluted and poured off and a solution of strong acid is added. This acid should contain 50 per cent chemically pure acid. Boil for five minutes. Wash the residue several times and finally fill the flask full of water and invert an annealing cup, Fig. 56, over it. Invert the flask, tapping to settle the gold. Remove the flask quickly, allowing the water from it to fall outside the annealing cup. Drain off the water remaining in the annealing cup, dry, brighten, and weigh. The gold and silver values in base bullion are always reported in ounces per ton, the same as ore.

Fig. 56.

Assay of Fine Bullion. Fine bullion is the term applied to bullion in which the gold and silver predominate. It is called gold bullion when there is more gold than silver, and silver bullion when the silver predominates. The percentages of gold and silver are expressed in fineness or parts in 1,000. Thus 50 per cent gold would be 500 fine.

Melting. Gold bullion is not received at the United States Assay Office unless it is 500 fine. The bar is weighed, stamped with a number, and goes to the melting room. Here it is melted and cast into a bar. This operation eliminates all possibility of fraud, as filling the centre of the bar with lead or copper. Borax and soda are generally used in melting, but when lead, copper, etc., are present, nitre is used to oxidize as much as possible of these metals. The charge is poured into a mold, and when cool the bar is taken out, cleaned with sand, and marked with its number. The slag is crushed in a mortar and panned in a gold pan, Fig. 57. Any shots of gold found in the slag are dried, and placed in an envelope bearing the same number as the bar. These shots and the sample clips are saved and melted together every four months and sent to the mint. This is called a "clip" bar.

Mass melts are made in order to have the gold in better form for shipment, the bars weighing a little

FIG. 57.

less than 1,000 ounces. These melts are made every Friday and shipped Saturday. The clips and shots from this bar go with it to the mint.

Sampling. Samples are cut from diagonally opposite corners of the bark, flattened in a diamond mortar, Fig. 58, and rolled into sheets in cornet rolls, Fig. 59. These sheets are cut into small pieces with shears, and are now ready for weighing. In the United States Assay Office there are two assayers; each takes a sample and assays it. Their results must check very closely.

FIG. 58.

Preliminary Assay. In order to determine approximately the composition of the bullion, so as to make up the proper proof, a preliminary assay is made. A proof assay is then made up to correspond with the composition as determined by the preliminary assay and is cupelled with the regular assay.

The losses in weight of the gold and silver contents of the proof determine the correction to be added to the results of the regular assay.

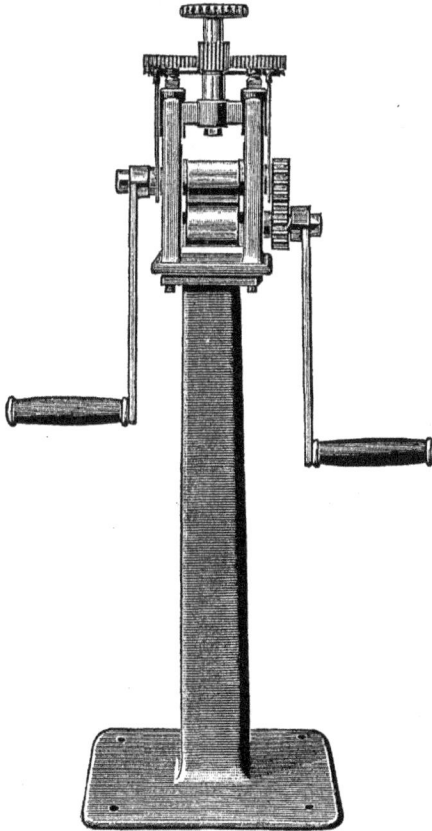

Fig. 59.

Weigh out 500 milligrams of bullion on the button balance, wrap in five grams of pure lead foil, and cupel

as in the assay of base bullion. In the United States
Assay Office the weights are marked double their
value, 500 milligrams being marked 1,000 and so on.
This saves multiplying everything by two.

The weight of the bullion taken, less the weight of
the gold and silver button, is considered base metals
and is represented in the proof by copper.

Suppose the preliminary button weighed 740, then:

$$1,000 - 740 = 260 \text{ base metals.}$$

If the gold in this button weighs 670 we have:

$$740 - 670 = 70 = \text{silver.}$$

We then have:

Base metal................ 260
Silver 70
Gold 670
                          ——
Total...............1,000

To part easily the button should contain twice as
much silver as gold, and if it does not contain the
required amount, it must be added in the assay and
also in the proof.

Then in this case the amount of silver to be added
is:

$$(670 \times 2) - 70 = 1,270,$$

and we have the proof made up as follows:

Gold 670
Silver (1,270+70) 1,340
Copper 260

Cupelling. The assay and proof are cupelled together, the cupels showing feathers. After cupellation the buttons weigh:

Assay 2,006

Proof 2,002

$(1,340+670)-2,002=8$, parts gold and silver lost in the proof, and the correct weight of the gold and silver is $2,006+8=2,014$.

Parting. The buttons are flattened into cornets by passing through a small pair of rolls, and are then stamped with the number of the assay. They are then placed in a platinum crate divided into sixteen compartments, each compartment holding a platinum capsule, Fig. 60, into which the cornet is placed. This crate containing the cornets is then placed into a bath of boiling nitric acid, of 52° Baumé and allowed to remain for ten minutes. It is then removed and placed in a bath of fresh acid, also boiling, for an additional ten minutes. The crate is then lifted out and washed thoroughly with distilled water, and dried. When thoroughly dry it is placed in the muffle and the cornets annealed at a low, red heat. They are now ready for weighing; they retain the numbers stamped on them and cannot become mixed.

Fig. 60.

Suppose after parting the cornets weigh:

Assay 671

Proof 668

Loss of gold in proof, $670-668=2$.

Corrected weight of the gold:

$$671 + 2 = 673,$$

and the corrected weight of the silver is:

$$2,014 - (673 + 1,270) = 71.$$

Then the corrected weight of the base metal is:

$$1,000 - (673 + 71) = 256$$

We then have:

Base metal................ 256
Silver   71
Gold 673
                          _____
Total................1,000

In the Helena Office the preliminary assay is made by touchstone. The needles range from 500 to 1,000 fine and are twenty points apart. The sample and one of the needles are rubbed on the stone, and the marks compared. This is repeated until the two marks correspond; the fineness of the bar being approximately the same as the corresponding needle. One proof is made up which is 900 fine and each assayer makes seven assays with this proof. The correction is then calculated for each bar in proportion to its fineness.

The assay certificate in use in the United States Assay Office is of the form shown in Fig. 61. An example is filled in.

U.S. MINT SERVICE—Form No. 42 D.

5 x 12¾.

No. ___13_0_0___

MEMORANDUM OF GOLD BULLION *deposited at the* UNITED STATES ASSAY OFFICE *at* HELENA, MONT.,

the ___13th___ day of ___June___, 190 4, by ___John Jones___

DESCRIPTION OF BULLION.	WEIGHT.			GOLD.			SILVER.			CHARGES FOR MELTING, REFINING PARTING, AND ALLOY.		NET VALUE.		RETURNED TO DEPOSITOR.			
	BEFORE MELTING.	AFTER MELTING AND DEDUCTIONS.												GOLD BAR.		COIN.	
	Ounces.	Dec.	Ounces.	Dec.	FINENESS 1000ths	VALUE. Dollars.	Cents	FINENESS Value @ $1.29 per St. Oz. 1000ths	Dollars.	Cents	Dollars.	Cents	Dollars.	Cents	Dollars.	Cents	Dollars.	Cents
Retort	4	77	4	34	637	55	87	265	1	16	1	22	54	19			54	39

FIG. 61.

Assayer in Charge.

VII

METHODS OF HANDLING WORK

Working by Numbers. A beginner in assaying should always number his crucibles, beginning at the left end of the row and numbering to the right. The holes of the mold and the spaces on the cupel and crucible trays should also be numbered. Keel is generally used for marking crucibles as it will not burn off. Chalk may be used for marking molds and trays.

Working by Position. After the assayer has worked by numbers for some time, the positions of the crucibles, cupels, etc., become so familiar to him that the numbers are no longer necessary and he can pick out an assay anywhere in the process and tell what it is.

Each sack should be marked with the date, name of the owner, character of the ore, and the determinations to be made on the sample. If a sulphide, the amount of nitre to be used is placed on the sack after the preliminary assay has been made.

Letters may be used instead of the full word when marking samples, as Q for quartz, O for oxide, C for carbonate, and S for sulphide, etc.

4-16 Geo. Jones; Au, Ag, Pb; S. N$=8$

The sacks are placed in a row on the left-hand side of the pulp balance, and as the proper charge of ore is weighed out, each sack is transferred, in turn, to the right-hand side of the balance. Before weighing

May 28, 1904	No.	No.	Ag	Au	Ag	Pb
Keating Sulphide.....	111	1	—		—	—
Keating Sulphide.....	111	2	—		—	—
Keating Sulphide.....	112	3	—		—	—
Keating Sulphide.....	112	4	—		—	—
Providence M. Co., S.	1	5				
Providence M. Co., S.	2	6				
Providence M. Co., S.	3	7				
Providence M. Co., S.	4	8				
Chris. Voelker, O.....	1	9				—
Chris. Voelker, O.....	2	10				—
Chris. Voelker, O.... .	3	11				—
Tregoning Sulphide...	61	12	—		—	—
Tregoning Sulphide...	61	13	—		—	—
Tregoning Sulphide...	62	14	—		—	—
Tregoning Sulphide...	61	15	—		—	—
Geo. Gould, O.......		16				—

Fig. 62

a tabulation is made, Fig. 62, which shows which ores are sulphides. The blank spaces show which are to be assayed for lead, etc.

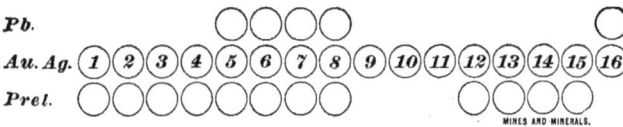

Fig. 63.

In weighing, the preliminary crucible is placed in front of the crucible containing the gold and silver assay and the lead crucible at the back, Fig. 63.

When all the crucibles are charged, the preliminaries, leads, and the oxidized ores are mixed and
covered, leaving only the sulphide ores uncovered,
Fig. 64.

After the preliminaries are made, the proper

FIG. 64.

amounts of nitre are charged into the crucibles
containing the sulphide ores, which are then mixed.

The first sample weighed is, of course, No. 1.
This is placed in the left-hand crucible. Its preliminary is then weighed and placed in the
crucible in front of it. No. 2's preliminary
is then weighed and placed in the second
preliminary crucible. Then No. 2 gold and
silver charge is weighed, and so on.

When the samples are all weighed, the

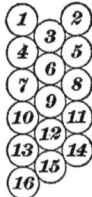

MINES AND MINERALS.
FIG. 65.

crucibles are placed in the muffle in the
order of Fig. 65, and are poured as shown
in Fig. 66, No. 16 being poured first,
then 15, 14, 13, and so on, just opposite to the order in which they were
put in.

The buttons are slagged, beginning
with the sixteenth, which will be the
coolest, and placed on the cupel tray
in the same order as Fig. 66.

The cupels are placed in the muf-
fle as in Fig. 67, and the buttons put into them in

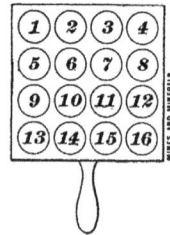

FIG. 66.

the order shown. When they are taken from the muffle they are placed on the tray in their original order, Fig. 66. The buttons are removed from the cupels and arranged on the base of the balance in the same order as the crucibles were at first, Fig. 63.

After weighing, the buttons are placed in porcelain crucibles, which are placed in an iron rack, Fig. 47, in the same order as Fig. 66. This tray should be made of iron one-eighth inch thick, with legs one-half inch long. It can be placed on the hot plate and the parting, washing, and drying done without removing it. It can then be transferred to the muffle and the gold brightened.

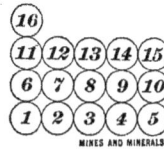

FIG. 67.

A convenient apparatus for washing may be made by placing an ordinary five-pint acid bottle on a shelf about two feet above the hot plate and siphoning the water, using a rubber tube and pinch-cock. A convenient form of record is an ordinary roll book, Fig. 68.

DESCRIPTION	No.	Au	Ag	Pb	Cu	Fe	SiO_2	Zn		

FIG. 68

Figure 69 shows the usual form of assay report.

ASSAYER AND CHEMIST

Helena, Montana, ...190

The samples of ore assayed for . . . .*contain.*

No.	Description	Ounces Per Ton Gold	Silver	Val. of Gold at $20 Per Oz.	Val. of Silver at . c. Per Oz.	Per Cent Lead	Per Cent Copper	Per Cent	Value Lead at Smelter Price	Value Copper at. . . .c.lb.	Total Value Per Ton

Charges, $. . . . .

. . . .*Assayer.*

FIG. 69.

VIII

LABORATORY TESTS

In making tests to determine the adaptability of a process to an ore it is necessary to sample the ore as described in "sampling for tests."

It is desired to know the amount of metal that will be extracted by a certain process, therefore some convenient fraction of a ton should be taken. As small a quantity as one assay ton is sometimes taken, but it is better to use four or more assay tons. The weight, in milligrams of the metal extracted, divided by the number of assay tons taken, will give the amount extracted per ton.

Amalgamation. Weigh out 5 A. T. of ore and place in an ordinary iron mortar or a Bucks amalgamating mortar with enough water to make a pulp, which should be thick enough to allow the globules of mercury to remain suspended in it. Add about $\frac{1}{2}$ gram potassium cyanide, $\frac{1}{10}$ A. T. salt, and $\frac{1}{2}$ A. T. mercury. Grind for several hours; then thin the pulp and stir to collect the mercury. Transfer the contents of the mortar to a gold pan and pan the ore into a second pan. Stir for a few minutes and pan again, saving the tailings. The mercury removed by the two pannings is placed in a porcelain dish

69

and particles of dirt are washed off. The amalgam is generally separated from the mercury by straining through a chamois or buckskin. Pour the mercury into skin, holding it so as to form a bag. Grasp the skin above the mercury with the left hand and twist with the right, allowing the strained mercury to fall into a porcelain dish. On opening the skin a lump of amalgam is found which is placed in an iron dish and heated to redness to expel the mercury. (Amalgam should never be heated in a muffle as the fumes of mercury are likely to carry some values which may salt the muffle and cause future assays to run high.) Wrap the bullion in sheet lead, cupel, weigh, and part as in an ordinary assay. The tailings are dried, sampled, and assayed. The difference between the assays of the "heads" and "tails" shows the amount extracted and should agree very closely with the bullion assay.

Suppose the heads assayed .50 ounce per ton gold; 11.0 ounces per ton silver; and the tails assayed .05 ounce per ton gold; .9 ounce per ton silver. Extraction is, therefore, .45 ounce per ton gold; 10.1 ounces per ton silver.

Five A. T. treated as above yielded 2.225 milligrams gold, 50.45 milligrams silver, which divided by 5 gives .445 ounce per ton and 10.09 ounces per ton.

Then we have practically extracted:

$$.50: \ 0.445 = 100: x = 89 \text{ per cent of the gold value.}$$
$$11.0 \ : 10.09 \ = 100: x = 91.7 \text{ per cent of the silver value.}$$

Amalgamation tests are sometimes made in bottles. The ore, mercury, and salt are placed in a heavy bottle with enough water to form a thin pulp. The bottle is corked and shaken for several hours. The contents are then panned and the mercury recovered, as in the mortar test.

A method recommended by Furman as especially adapted to silver ores is as follows: Place the finely ground ore in a copper pan, the inside of which has been previously amalgamated, with enough water to make a thin pulp. Stir with a stick for several hours and pan off the tailings. The amalgam will adhere to the pan and is scraped off and tested for the recovery of the gold and silver as in the first process.

An amalgamated copper pan may be used to pan the pulp from a mortar test.

Gold amalgam may be parted with nitric acid the same as gold and silver buttons of an ordinary assay.

A little sodium amalgam will prevent the mercury flowing. To prepare sodium amalgam, about two ounces of mercury are placed in a heavy china cup and the sodium introduced on the end of an iron wire about twelve inches long. A piece of sodium about the size of a pea is placed on the end of the wire and brought in contact with the mercury. The union takes place with an explosion and a brilliant flash. The operator should stand back as far as possible and protect his eyes.

Chlorination. The roasted ore after being passed through a 70-mesh sieve is placed in a bottle, moistened with water, and a stream of chlorine gas introduced at the bottom.

Bleaching powder and sulphuric acid are introduced into a generator and a constant supply of chlorine is given off from their union. Chlorine is passed into the test until the bottle is full and the gas escapes at the top, as will be detected by its odor. The tube is now removed; the bottle corked and allowed to stand for twenty-four hours in a warm place. Boiling water is now added to dissolve all soluble salts, and the solution decanted through a filter. Warm the solution to expel free chlorine. Precipitate the gold with ferrous sulphate and filter. Burn the filter, scorify with test lead, cupel, part, and weigh in the usual manner.

The roasted ore may be placed in a heavy bottle with enough water to make a thin pulp. Bleaching powder and a short test tube filled with sulphuric acid are placed in the bottle, after which it is tightly corked. The cork should be covered with lead foil and wired on, as the pressure of the gas is very great. The bottle is inverted and its contents mixed by shaking. Allow it to stand twelve hours in a warm place, wash and precipitate as described above. Another method of precipitation is to boil the solution with iron filings. Decant through a filter and wash the filings, after which they are dissolved in sulphuric acid and the solution filtered. Gold precipitated by this method is easy to filter.

Cyanide. A preliminary test should always be made to determine the adaptability of the ore to the cyanide process. This test can be made in an ordinary glass funnel, the tube of which is fitted with a rubber tube and pinch-cock. Ordinary filter paper is used.

Weigh out 2 A. T. of ore which has been passed through a 60-mesh screen and place on the filter. Weigh out ½ gram of cyanide and dissolve in 100 cubic centimetres of water. Add the solution slowly, allow to stand for twenty-four hours, remove the pinch-cock and allow the solution to drain off. Wash three times, allowing the wash water to remain in contact with the ore for two hours each time. Dry and assay the tailings. The difference between the assay of the heads and the tails will be the amount extracted. The filtrate should be titrated with silver nitrate to determine the consumption of cyanide.

This test will generally show whether or not an ore will cyanide, but its results are not final, as some ores require preliminary washing, with lime water or other chemical washes, before the cyanide will attack the gold. Sometimes washing with water alone will extract the undesirable constituents of the ore. Working tests can be made in bottles, the bottoms of which have been broken off. They are inverted and filled with a rubber stopper, into which is inserted a glass tube with a rubber tube and pinch-cock on the lower end.

A wooden grid is made to fit the bottle, and on this is placed coarse gravel, on which are placed, in turn, finer gravel and sand, and a filter cloth of canvas completes the filter bed.

If exhaustive tests are to be made, six or more of these bottles will be needed.

The following points should be investigated to determine the adaptability of an ore to cyanide treatment:

1. — Percolation.
2. — Extraction.
3. — Consumption of cyanide.
4. — Strength of solution required.
5. — Time required.
6. — Precipitation.

Five hundred grams of ore for each test is a convenient quantity to use and should be passed through a 40 to 60-mesh screen, the size of the screen depending on the porosity of the ore. The more porous the ore the coarser the size necessary.

An absorption test may be made as follows: Weigh out 100 grams of ore and place in a dry beaker about three inches in diameter. Place the beaker containing the ore under a burette filled with water, so that the drop will fall in the centre of the ore. Allow the water to drop very slowly. When the moisture has reached the sides of the beaker, the operation is finished and the number of cubic centimetres of water used is the percentage of absorption. Now allow more water to drop in until the ore is thoroughly wetted. This amount shows the percentage of saturation. If the absorption was 25 cubic centimetres, and it took 5 cubic centimetres additional to saturate the ore, then the absorption is 25 per cent and the saturation 30 per cent.

Acidity. Ten grams of ore should be agitated with water and tested for acidity with litmus paper. If it shows the presence of acids, the test should be treated for two hours with a saturated solution of lime. The lime water is then drawn off and the ore leached with cyanide.

Leaching. Cut out a sample of the ore for assay, marking it "heads." Weigh out 500 grams of ore for each test and place in the bottles. If the ore shows acidity, wash with lime water.

If the ore has been washed, allowance must be made for absorption when making up solutions.

If 500 grams of ore are used and the absorption is 25 per cent, the ore has absorbed 125 cubic centimetres of water. Suppose we wish to use 500 cubic centimetres of solution .10 per cent cyanide. Weigh out 500 milligrams of cyanide and dissolve in 375 cubic centimetres of water, which, when added to the wet ore, will contain .10 per cent cyanide. A 200-cubic-centimetre measuring cylinder will be found very convenient.

Make up six solutions containing respectively .10, .15, .20, .25, .30, and .35 per cent cyanide. Or, if the preliminary tests show a great consumption of cyanide, a stronger series may be used, .20, .30, .35, .40, .45, .50.

Charge No. 1 test with the weakest solution, No. 2 with the next, etc. Allow to stand for twenty-four hours and then drain, note the time for the solution to drain off, which will show the adaptability of the ore to percolation. Wash several times, allowing the wash water to remain in contact with the ore about two hours each time. Dry and assay the tailings. The difference between the assay of the "heads" and "tails" will be the extraction. Dilute the solution to 1,000 cubic centimetres and evaporate 100 cubic centimetres of it to dryness, with 30 grams litharge and assay. The result multiplied by 10 will give the extraction.

Precipitation. Fill a long tube with clean zinc shavings and allow 100 cubic centimetres of the solution to drop slowly through the zinc. If the zinc darkens at the top the precipitation is good, but if it becomes black near the lower end, the conditions are unfavorable. After all the solution has passed through, wash the zinc and evaporate the solution to dryness with 30 grams litharge, and assay. The difference between this assay and the first will give the amount precipitated. If the extraction is not complete after twenty-four hours, another series of tests should be made, allowing the ore to leach for forty-eight hours.

Silver Nitrate Test. After precipitation with zinc the solution should be tested for cyanide in order to know the amount of cyanide consumed. Dissolve 13.06 grams of chemically pure silver nitrate in 1,000 cubic centimetres distilled water. Every .1 cubic centimetre of this solution added to 10 cubic centimetres of cyanide solution represents .01 per cent cyanide.

Take 10 cubic centimetres of cyanide solution in a test tube and allow the silver nitrate solution to

drop slowly into it from a burette. After every two
or three drops shake the tube to dissolve the precipitated silver cyanide. When the precipitate will no
longer dissolve, the operation is ended, and the
reading of the burette will be the amount of cyanide
present. This amount subtracted from the original

Fig. 70.

strength of solution will give the amount of cyanide
consumed in the test.

Fig. 70 shows a laboratory cyanide outfit manufactured by F. W. Braun and Company, Los Angeles,
California.

Concentration. Crush five pounds of the ore to be
tested through a 10-mesh screen. Screen through a

20-mesh screen and then a 40-mesh. We then have three sizes: 10, 20, and 40 mesh. Cut out a sample from each of these sizes for assay.

Weigh out one pound of each size, pan very carefully, saving the tailings. Pan the tailings over again, and if any concentrates are obtained add them to the first concentrates. Allow the tailings to settle and pour off the water. Dry and assay the tailings and concentrates.

The assays of the ore and concentrates will indicate the degree of concentration, while the assay of the tailings will show the loss. The results with the different sizes will show from which size the best results may be expected. However, in mill work the ore is crushed as coarse as consistent with good saving, as fine crushing produces a large amount of slimes, which carry values into the tailings.

COMPARISONS AND EQUIVALENTS

TROY WEIGHT

For weighing precious metals, such as Gold, Platinum, Silver, etc., Troy weights are used exclusively. The U. S. Standard Troy pound, was copied in 1827 from the imperial Troy pound of England for the use of the United States Mint, and there deposited. It is standard in air at 62° Fahr., the barometer at 30 inches.

24 grains = 1 pwt.
480 " = 20 " = 1 oz.
5760 " = 240 " = 12 " = 1 lb. = 22.816 cu. in. of distilled water at 62° Fahr.

AVOIRDUPOIS WEIGHT

For weighing base metals such as Lead, Antimony, Tin, etc., and the weight in general use in trade and commerce in the U. S.

1 drachm = 27.34375 grains Troy.
16 " = 1 oz. = 437.5 " " "
256 " = 16 " = 1 lb. = 1.2153 lb. Troy.
6400 " = 400 " = 25 " = 1 quarter.
25600 " = 1600 " = 100 " = 4 " = 1 cwt.
512000 " = 32000 " = 2000 " = 80 " = 20 " = 1 ton.

APOTHECARES' WEIGHT

20 grains = 1 scruple.
60 " = 3 " = 1 drachm.
480 " = 24 " = 8 " = 1 oz.
5760 " = 288 " = 96 " = 12 " = 1 lb.

METRIC, OR FRENCH WEIGHTS

	Grams	Troy grs.	Troy ozs.	Troy lbs.	Avoir. ozs.	Avoir. lbs.
1 Milligram	.001	.01543				
1 Centigram	.01	.15432				
1 Decigram	.1	1.5432				
1 Gram	1.	15.432	.032	.00267	.03528	.0022047
1 Decagram	10.		.321	.02679	.3528	.022046
1 Hectogram	100.		3.215	.26792	3.52758	.22046
1 Kilogram	1000.		32.150	2.6792	35.2758	2.2046
1 Myriagram	10000.			26.792		22.046
1 Quintal	100000.			267.92		220.46
1 Tonneau	1000000.			2679.2		2204.6

The unit of the metric system is the gram =15.438395 Troy grains, or the weight of 1 c. c. of distilled water at 4° C.

ASSAY TON WEIGHTS

The Assay Ton Weights is a system made up from a comparison of the Avoirdupois, Troy and Gram Weights, and will be found extremely simple and useful, saving a vast amount of calculation and labor.

The Unit of the System is the Assay Ton (A. T.) =29.166 grams. Its derivation will be seen at a glance.

1 lb. Avoirdupois =7,000 Troy grains.
2,000 lbs. =1 ton.
2,000×7,000 =14,000,000 Troy grains, in one ton Avoirdupois.
480 Troy grains =1 oz. Troy.
14,000,000÷480 =29,166 Troy ozs. in 2,000 lbs. Avoirdupois.
There are 29.166 milligrams in one assay ton (A.T.); hence
2,000 lbs. is to 1 A. T., as 1 oz. Troy is to 1 milligram. Therefore if 1 A. T. of ore assays 1 milligram of gold or silver, the ton contains 1 ounce Troy.

COMPARISON OF AVOIRDUPOIS, MET

	AVOIR. OUNCE	AVOIR. POUND	MILLIGRAMS	CENTIGRAMS	DECAGRAMS	GRAMS
1 Avoir. Oz....	1.	.06250	28349.5403	2831.95403	283.495403	28.3495403
1 " Lb....	16.	1.	453592.6449	45359.26449	4535.926449	453.5926449
1 Milligram ..	.00003527394	.00000220462	1.	.1	.01	.001
1 Centi "	.0003527394	.0000220462	10.	1.	.1	.01
1 Deci "	.003527394	.000220462	100.	10.	1.	.1
1 Gram	.03527394	.00220462	1000.	100.	10.	1.
1 Decagram ..	.3527394	.0220462	10000.	1000.	100.	10.
1 Hecta "	3.527394	.220462	100000.	10000.	1000.	100.
1 Kilo "	35.27394	2.20462	1000000.	100000.	10000.	1000.
1 Troy Grain .	.00228571	.000142857	64.79897	6.479897	.6479897	.06479897
1 ' Pwt...	.0548571	.0034285	1555.1754	155.51754	15.551754	1.551754
1 " Ounce.	1.0971428	.0685714	31103.495	3110.3495	311.03495	31.103495
1 " Pound.	13.165714	.822857	373241.9478	37324.19478	3732.419478	373.2419478
1 Assay Ton ..	1.0288232					29.166666

METRIC, TROY AND ASSAY TON WEIGHTS

DECAGRAMS	HECTAGRAMS	KILOGRAMS	TROY GRAIN	TROY PWT.	TROY OUNCE	TROY POUND
2.83495403	.283495403	.0283495403	437.5	18.22917	.911458	.07595485
45.35926449	4.535926449	.4535926449	7000.	291.66666	14.583333	1.215277
.0001	.00001	.000001	.0154322349	.0006430145	.000032150727	.00000267922725
.001	.0001	.00001	.154322349	.006430145	.00032150727	.0000267922725
.01	.001	.0001	1.54322349	.06430145	.0032150727	.000267922725
.1	.01	.001	15.4322349	.6430145	.032150727	.00267922725
1.	.1	.01	154.322349	6.430145	.32150727	.0267922725
10.	1.	.1	1543.22349	64.30145	3.2150727	.267922725
100.	10.	1.	15432.349	643.0145	32.150727	2.67922725
.006479897	.0006479897	.00006479897	1.	.041666	.0020833	.000173611
.15551754	.015551754	.0015551754	24.	1.	.05	.0041666
3.1103495	.31103495	.031103495	480.	20.	1.	.08333333
37.32419478	3.732419478	.3732419478	5760.	240.	12.	1.

TABLE TO CONVERT METRIC WEIGHTS INTO AVOIRDUPOIS AND TROY WEIGHTS

As one gram is equal to 15.432 × grains, or .03527 Avoirdupois ounce, or .03215 Troy ounce, to convert:

Grams	into grains	multiply by	15.432
Centigrams	" grains	"	0.15432
Milligrams	" grains	"	0.01543
Kilograms	" Avoirdupois ounces. . .	"	35.2739
Grams	" Avoirdupois ounces. . .	"	.03527
Kilograms	" Avoirdupois pounds . .	"	2.2046
Kilograms	" Troy ounces.	"	32.1507
Grams	" Troy ounces.	"	.03215

TABLE TO CONVERT AVOIRDUPOIS AND TROY WEIGHTS INTO METRIC WEIGHTS

As one grain is equal to 0.0648 and one Avoirdupois ounce is equal to 28.3495 grams, and one Troy ounce is equal to 31.1035 grams, to convert:

Grains	into grams	multiply by	0.0648
Grains.	" centigrams	"	6.4799
Grains.	" milligrams	"	64.799
Avoirdupois ounces. . .	" kilograms	"	0.02835
Avoirdupois ounces. . .	" grams	"	28.3495
Avoirdupois pounds. . .	" kilograms	"	0.4536
Troy ounces.	" kilograms	"	0.0311
Troy ounces.	" grams	"	31.1035

LINEAR MEASURE, U. S. STANDARD

The Standard unit of the United States and British linear measure is the yard. It was intended to be exactly the same for both countries, but in reality the United States' yard exceeds the British standard by .00087 inch. The actual standard of length for the United States is a brass scale 82 inches long prepared for the Coast Survey and deposited in the office of Weights and Measures at the U. S. Treasury Department, Washington. The yard is between the 27th and 63d inches of this scale. The temperature at which this scale is designed to be standard, and at which it is used in the U. S. Coast Survey is 62° Fahrenheit.

Inches.	Foot.	Yard.	Fathom.	Perch.	Furlong.	Mile.	League.
12	1.						
36	3.	1.					
72	6.	2.	1.				
198	16.5	5.5	2.75	1			
7920	660.	220.	110.	40	1		
63360	5280.	1760.	880.	320	8	1	
190080	15840.	5280.	2640.	960	24	3	1

LINEAR MEASURE, METRIC

	Metre.	U. S. Ins.	Feet.	Yards.	Miles.
1 Millimetre	.001	.03937	.00328		
1 Centimetre	.01	.3937	.0328		
1 Decimetre	.1	3.937	.32808		
1 Metre	1.	39.3704	3.2808	1.0936	
1 Decametre	10.	393.704	32.808	10.936	
1 Hectometre	100.		328.08	109.36	.0621375
1 Kilometre	1000.		3280.8	1093.6	.621375
1 Myriametre	10000.		32808.	10936.	6.21375

TABLE TO CONVERT U. S. LINEAR MEASURE INTO METRIC LINEAR MEASURE

As one inch is equal to 0.0254 metres; to convert:

Inches.........into metres......multiply by 0.0254
"   .... " centimetres...   "   "   2.5399
"   .... " millimetres ...   "   "   25.3997

TABLE TO CONVERT METRIC LINEAR MEASURE INTO U. S. LINEAR MEASURE

As one metre is equal to 39.370 inches; to convert:

Metres.........into inches.......multiply by 39.370
Centimetres.... "   " ....   "   "   0.3937
Millimetres..... "   " ....   "   "   0.03937

CUBIC MEASURE, U. S. STANDARD

1,728 cubic inches =   1 cubic foot.
46,656 cubic inches = 27 cubic feet = 1 cubic yard.

A cubic foot of water weighs 62½ pounds, and contains 1,728 cubic inches, or 7½ gallons.

CUBIC MEASURE, METRIC

	Cu. Metres		U. S. Cu. Ins.		U. S. Cu. Ft.		U. S. Cu. Yds.
1 Cubic Centimetre	.000001	=	.061025				
1 Cubic Decimetre	.001	=	61.025				
1 Centistere	.01	=	610.25	=	.353156	=	.13080
1 Decistere	.1	=	6102.5	=	3.53156	=	1.3080
1 Stere	1.	=	....	=	35.3156	=	13.080
1 Decastere	10.	=	....	=	353.156	=	130.80
1 Hectostere	100.	=	....	=	3531.56		

SQUARE MEASURE, U. S. STANDARD

Inches		Foot.		Yard.		Perch.		Rood.		Acre.
144	=	1.								
1296	=	9.	=	1.						
39204	=	272.25	=	30.25	=	1				
1568160	=	10890.	=	1210.	=	40	=	1		
6272640	=	43560.	=	4840.	=	160	=	4	=	1

An acre is 69.5701 yards square; or 208.710321 feet square.

SQUARE MEASURE, METRIC

	Sq. Metres.		U. S. Sq. In.		Sq. Feet.		Sq. Yards.		Acres.
1 Sq. Centimetre	.0001	=	.155	=	.10764	=	.01196		
1 Sq. Decimetre	.01	=	15.5	=	10.764	=	1.196		
1 Centiare	1.	=	1550.03	=	1076.4	=	119.6	=	.00024
1 Are	100.	=	155003.	=	10764.	=	11960.	=	.0247
1 Hectare	10000.	=	....	=	107641.			=	2.47

DRY MEASURE, U. S. STANDARD

Bushel.	Pecks.	Gallons.	Quarts.	Pints.	
1 = 4	= 8	= 32	= 64		Cu. Ins.
	1 = 2	= 8	= 16	=	537.6
		1 = 4	= 8	=	268.8
			1 = 2	=	67.2
			1	=	33.6

NOTE. — The standard U. S. bushel is the Winchester bushel, which is in cylinder form, $18\frac{1}{2}$ inches diameter and 8 inches deep, and contains 2,150.42 cubic inches.

The English Imperial bushel .... = $\begin{cases} 2218.192 \text{ cubic inches.} \\ 1.03152 \text{ U. S. bushels.} \end{cases}$

The English Quarter........... = $\begin{cases} 8 \text{ Imperial bushels.} \\ 8\frac{1}{4} \text{ (nearly) U. S. bushels.} \\ 10.2694 \text{ cubic feet.} \end{cases}$

DRY MEASURE, METRIC

(IN THE METRIC SYSTEM THE DRY AND LIQUID MEASURES ARE THE SAME.)

	Litres.		U. S. Cu. Ins.		U. S. dry.
1 Millilitre ...... =	.001	=	.061	=	.0018 pint.
1 Centilitre ..... =	.01	=	.61	=	.018 pint.
1 Decilitre ...... =	.1	=	6.1	=	.18 pint.
1 Litre.......... =	1.	=	61.02	=	1.8 pints.
1 Decalitre...... =	10.	=	610.16	=	9.08 quarts.

			U. S. Cu. Ft.		
1 Hectolitre..... =	100.	=	3.531	=	2.837 bushels.
1 Kilolitre ...... =	1000.	=	35.31	=	28.378 bushels.
Myrialitre..... =	10000.	=	353.1	=	283.7 bushels.

LIQUID OR WINE MEASURE, U. S. STANDARD

NOTE. — The standard unit of Liquid Measure adopted by the U. S. Government is the Winchester Wine Gallon, which contains 231 cubic inches, and holds 8.339 pounds Avoirdupois of distilled water, at its maximum density weighed in air, the barometer being at 30 inches.

Gallons.	Quarts.	Pints.	Gills.	Cu. Ins.	Cu. C. M.
1 =	4 =	8 =	32	= 231.	= 3785.00
	1 =	2 =	8	= 57.75	= 946.22
		1 =	4	= 28.875	= 473.11

A gallon of water (U. S. Standard) weighs $8\frac{1}{3}$ pounds, and contains 231 cubic inches.

DIMENSIONS OF CYLINDERS, HOLDING APPROXI-
MATELY BELOW NAMED, U. S. STANDARD MEASURES

Dia. Height.

A cylinder $1\frac{3}{4}$ — 3 contains approximately 1 gill U. S. standard.

"	$2\frac{1}{4}$ — $3\frac{3}{8}$	"	"	$\frac{1}{2}$ pint	"	"
"	$3\frac{1}{2}$ — 3	"	"	1 pint	"	"
"	$3\frac{1}{2}$ — 6	"	"	1 quart	"	"
"	7 — 6	"	"	1 gall.	"	"
"	14 — 12	"	"	8 galls.	"	"
"	14 — 15	"	"	10 galls.	"	"

LIQUID OR WINE MEASURE, METRIC

In the Metric system the liquid and dry measures are the same.

	Litres.		U. S. Cu. Ins.		U. S.
1 Millilitre.........	.001	=	.061	=	.00845 gill.
1 Centilitre........	.01	=	.61	=	.0845 gill.
1 Decilitre.........	.1	=	6.1	=	.845 gill = .2113 pints.
1 Litre.............	1.	=	61.02	=	2.113 pints = 1.056 quarts.
1 Decalitre.........	10.	=	610.16	=	2.641 gallons.
			U. S. Cu. Ft.		
1 Hectolitre........	100.	=	3.531	=	26.417 gallons.
1 Kilolitre.........	1000.	=	35.31	=	264.17 gallons.
1 Myrialitre........	10000.	=	353.1	=	2641.7 gallons.

LIQUID MEASURE, APOTHECARY

Gallons.		Pints.		Ounces.		Drams.		Mins.		Cu. Ins.		Grains of Water.		Cu. C. M.
1	=	8	=	128	=	1024	=	61440	=	231.	=	58328.886	=	3785.00
		1	=	16	=	128	=	7680	=	28.875	=	7291.1107	=	473.11
				1	=	16	=	480	=	1.8047	=	455.6944	=	29.57
						1	=	60	=	0.2256	=	56.9618	=	3.70

TABLE TO CONVERT U. S. LIQUID MEASURE INTO METRIC LIQUID MEASURE

As one U. S. liquid ounce is equal to 29.572 cubic centimetres and one pint is equal to 0.4731 litres; to convert:

Pints............	into litres............	.multiply by 0.4731
Fluid ounces......	into cubic centimetres......	" 29.572

TABLE TO CONVERT METRIC LIQUID MEASURE INTO U. S. LIQUID MEASURE

As one Litre, or 1000 cc., is equal to 33.8149 fluid ounces; to convert:

Litres...........	into fluid ounces...........	.multiply by 33.815
Cubic centimetres	" "	" " 0.0338
Litres...........	" pints............	" " 2.113

COMPARISON OF VARIOUS LIQUID MEASURES

				Grains of Water @ 62° F.		Cubic Centimetre.
1 English........	Imperial Gallon	=	277.274 cu. inches	=	70,000.00	=	4,543.000
1 "	Wine or Winchester	"	231.000 " "	=	58,318.00	=	3,785.200
1 "	Corn..............	"	268.000 " "	=	67,861.00	=	4,402.900
1 "	Ale...............	"	282.000 " "	=	71,193.40	=	4,619.200

1 cu. ft................= 283.15 cc.
1 cu. inch..............= 16.38 cc.
0.061027 cu. inch.......= 1   cc.

EXPANSION OF METALS THROUGH HEAT

BODIES	EXPANSION IN 0° TO 100° C.	
	DECIMALS	FRACTIONS
Lead.............................	0.002848	1–351
Iron, wrought......................	0.001167	1–856
Iron, cast	0.001110	1–901
Glass, white.......................	0.000861	1–1161
Gold............................	0.001552	1–645
Copper..........................	0.001717	1–582
Marmor (Marble of Carrara)..........	0.000849	1–1178
Brass............................	0.001920	1–521
Platinum	0.000856	1–1167
Sandstone........................	0.001174	1–852
Silver............................	0.001909	1–524
Steel, hardened....................	0.001225	1–816
" soft........................	0.001079	1–926
Zinc, rolled.......................	0.003331	1–302
Zinc, cast........................	0.002987	1–336

TABLE OF HEATING DEGREES

		C.	F.
Red..........................commences at		525°	955°
Dark Red.......................	" "	700°	1292°
Cherry Red.....................	" "	850°	1562°
Light Red	" "	900°	1652°
Yellow.........................	" "	1100°	2012°
White..........................	" "	1300°	2372°
Full White......................	" "	1500°	2732°

PHYSICAL CONSTANTS OF VARIOUS METALS

METALS	SYMBOL	ATOMIC WEIGHT	SPECIFIC GRAVITY	SPECIFIC HEAT	MELTING POINTS		COLOR
					DEG. C	DEG. F	
Aluminum	Al	27.00	2.56	0.212	700	1292	Silver White
Antimony	Sb	120.00	6.71	0.051	432	809	Bluish White
Arsenic	As	74.90	5.67	0.081			Lead Gray
Barium	Ba	136.86	3.75	0.047	1200	2192	
Bismuth	Bi	207.50	9.80	0.031	258	496	Silver White
Cadmium	Cd	111.70	8.60	0.057	320	608	Tin White
Caeseum	Cs	132.70	1.88		26	78	
Calcium	Ca	39.91	1.57	0.170		Red heat	
Chromium	Cr	52.40	6.80	0.120			
Cobalt	Co	58.60	8.50	0.110	1600	2912	Steel Gray
Copper	Cu	63.20	8.82	0.094	1050	1922	Red
Didymium	Di	146.80	6.54	0.046			Yellowish
Glucinum	Gl	9.08	2.07	0.580			
Gold	Au	196.20	19.32	0.032	1102	2016	Yellow
Indium	In	113.40	7.42	0.057	176	348	
Iridium	Ir	192.50	22.42	0.033	2500	4532	
Iron	Fe	55.90	7.86	0.110	1600	2912	
Iron Cast					1530	2786	Gray
Iron Wro·ght					1808	3286	
Lanthanum	La	138.50	6.20	0.045			
Lead	Pb	206.40	11.37	0.031	322	612	Bluish
Lithium	Li	7.01	0.59	0.941	180	356	

Magnesium	Mg	23.94	1.74	0.250	750	1382	Grayish White
Manganese	Mn	54.80	8.00	0.120	1650	3002	Tin White
Mercury	Hg	199.80	13.59	0.032	40	104	Silver White
Molybdenum	Mo	95.90	8.60	0.072		2799	Silver White
Nickel	Ni	58.60	8.80	0.110	1537		
Niobium	Nb	94.00	6.27				
Osmium	Os	195.00	22.48	0.031	2500	4532	
Palladium	Pd	106.20	11.50	0.059	1500	2732	Whitish
Platinum	Pt	194.30	21.50	0.033	1808	3286	
Potassium	K	39.03	0.87	0.170	62	143	
Rhodium	Rh	104.10	12.10	0.058	2000	3632	
Rubidium	Rb	85.20	1.52	0.077	38	100	
Ruthenium	Ru	103.50	12.26	0.061	1800	3272	
Silver	Ag	107.66	10.53	0.056	1023	1873	White
Sodium	Na	22.99	0.97	0.290	95	203	Gray
Steel					1808	3286	
Strontium	Sr	87.20	2.54	0.074			
Tantalum	Ta	182.00	10.80				
Tellurium	Te	126.30	6.25	0.047	525	977	Dark Gray
Thallium	Tl	203.70	11.85	0.034	288	550	White
Thorium	Th	233.41	7.7	0.028			
Tin	Sn	117.40	7.29	0.056	228	442	
Titanium	Ti	48.00		0.130			
Tungsten	W	184.00	19.10	0.033			
Uranium	U	240.00	18.70	0.028			Gray
Vanadium	V	51.10	5.50				
Zinc	Zn	64.90	7.15	0.094	411	772	Bluish White
Zirconium	Zr	90.40	4.15	0.066			

TABLE FOR ESTIMATING THE VALUE OF SILVER PER TROY OUNCE AT DIFFERENT DEGREES OF FINENESS, BASED $.50 PER OUNCE 1000 FINE

To find the present market value of silver at any given time, add 2% for every 1c. above 50c.

Example: To estimate the value of 1 ounce of silver 750 fine, presuming the market value of silver is 63c. per ounce for silver 1000 fine, which is 13 cents, or 26% above the computed value in table below, thus: One ounce of silver 750 fine is worth 37½c.+26%=47¼c.

FINE	$	CENTS	FINE	$	CENTS	FINE	$	CENTS	FINE	$	CENTS
10		00.50	260		13	510		25.50	760		38
20		01	270		13.50	520		26	770		38.50
30		01.50	280		14	530		26.50	780		39
40		02	290		14.50	540		27	790		39.50
50		02.50	300		15	550		27.50	800		40
60		03	310		15.50	560		28	810		40.50
70		03.50	320		16	570		28.50	820		41
80		04	330		16.50	580		29	830		41.50
90		04.50	340		17	590		29.50	840		42
100		05	350		17.50	600		30	850		42.50
110		05.50	360		18	610		30.50	860		43
120		06	370		18.50	620		31	870		43.50
130		06.50	380		19	630		31.50	880		44
140		07	390		19.50	640		32	890		44.50
150		07.50	400		20	650		32.50	900		45
160		08	410		20.50	660		33	910		45.50
170		08.50	420		21	670		33.50	920		46
180		09	430		21.50	680		34	930		46.50
190		09.50	440		22	690		34.50	940		47
200		10	450		22.50	700		35	950		47.50
210		10.50	460		23	710		35.50	960		48
220		11	470		23.50	720		36	970		48.50
230		11.50	480		24	730		36.50	980		49
240		12	490		24.50	740		37	990		49.50
250		12.50	500		25	750		37.50	1000		50

THE VALUE OF GOLD PER TROY OUNCE AT DIFFERENT DEGREES OF FINENESS, BASED ON $20.6718 PER OUNCE FOR 1000 FINE

FINE	$	CENTS	FINE	$	CENTS	FINE	$	CENTS	FINE	$	CENTS
10		20.67	260	5	37.47	510	10	54.26	760	15	71.06
20		41.34	270	5	58.14	520	10	74.94	770	15	91.73
30		62.02	280	5	78.81	530	10	95.61	780	16	12.40
40		82.69	290	5	99.48	540	11	16.28	790	16	33.07
50	1	03.36	300	6	20.16	550	11	36.95	800	16	53.75
60	1	24.03	310	6	40.83	560	11	57.62	810	16	74.42
70	1	44.70	320	6	61.50	570	11	78.29	820	16	95.09
80	1	65.37	330	6	82.17	580	11	98.97	830	17	15.76
90	1	86.05	340	7	02.84	590	12	19.64	840	17	36.43
100	2	06.72	350	7	23.51	600	12	40.31	850	17	57.11
110	2	27.39	360	7	44.19	610	12	60.98	860	17	77.78
120	2	48.06	370	7	64.86	620	12	81.65	870	17	98.45
130	2	68.73	380	7	85.53	630	13	02.33	880	18	19.12
140	2	89.41	390	8	06.20	640	13	23.00	890	18	39.79
150	3	10.08	400	8	26.87	650	13	43.67	900	18	60.46
160	3	30.75	410	8	47.55	660	13	64.34	910	18	81.14
170	3	51.42	420	8	68.22	670	13	85.01	920	19	01.81
180	3	72.09	430	8	88.89	680	14	05.68	930	19	22.48
190	3	92.76	440	9	09.56	690	14	26.36	940	19	43.15
200	4	13.44	450	9	30.23	700	14	47.03	950	19	63.82
210	4	34.11	460	9	50.90	710	14	67.70	960	19	84.50
220	4	54.78	470	9	71.58	720	14	88.37	970	20	05.17
230	4	75.45	480	9	92.25	730	15	09.04	980	20	25.84
240	4	96.12	490	10	12.92	740	15	29.72	990	20	46.51
250	5	16.80	500	10	33.59	750	15	50.39	1000	20	67.18

PROSPECTOR'S GOLD TABLE

For Determining the value of Free Gold Per Ton (2,000 lbs.) of Quartz or Cubic Yard of Gravel

The table below furnishes an exceedingly simple method for determining the value of Free Gold in a ton of gold-bearing quartz, or a cubic yard of auriferous gravel.

Take a sample of four (4) pounds of quartz, pulverize it to the usual fineness for horning; wash it carefully by batea, pan, or other means; amalgamate the gold by the application of quick-silver; volatilize the quicksilver by blow-pipe or otherwise; weigh the resulting button, and the value given in the table opposite such weight will be the value in free gold per ton of 2,000 lbs. of quartz.

Example. — Sample of four lbs. produces button weighing one grain, the fineness of the gold being 830; then the value of one ton of such quartz will be $17.87.

If the sample of four pounds should produce a button weighing, say four tenths of a grain (.4) then the value of such quartz would be (830 fine) $7.14 per ton.

WEIGHT WASHED GOLD 4 LB. SAMPLE GRAINS	FINENESS 780 VALUE PER OZ. $16.12	FINENESS 830 VALUE PER OZ. $17.15	FINENESS 875 VALUE PER OZ. $18.08	FINENESS 920 VALUE PER OZ. $19.01
.1	$ 1.68	$ 1.78	$ 1.88	$ 1.98
.2	3.36	3.57	3.76	3.96
.3	5.03	5.36	5.65	5.94
.4	6.71	7.14	7.53	7.92
.5	8.40	8.93	9.42	9.90
.6	10.07	10.73	11.30	11.88
.7	11.75	12.51	13.19	13.86
.8	13.43	14.29	15.07	15.84
.9	15.11	16.08	16.95	17.82
1	16.79	17.87	18.84	19.81
2	33.59	35.74	37.68	39.62
3	50.38	53.61	56.52	59.43
4	67.18	71.49	75.36	79.24
5	83.97	89.36	94.20	99.05

Gold Value of a Cubic Yard of Gravel

To determine the gold value of a cubic yard of auriferous gravel, the same table can also be used.

Take a sample of sixty (60) pounds of gravel, pulverize it, and carefully wash it by batea, pan or otherwise; amalgamate the gold, volatilize the quicksilver; weigh the button, and in column in table, opposite the weight, will be found the gold value of the cubic yard of gravel.

Example. — Sample of sixty pounds produces button weighing one grain, the fineness of the gold being 780; then the value of one cubic yard of such gravel would be $1.67. This is arrived at by pointing off one point, or dividing the value given in the table by 10.

If the sample of sixty pounds yields a button weighing five tenths (.5) of a grain, then the value of the gravel would be — gold being 780 fine — $0.84 per cubic yard.

Simple Ore Tests

The following simple tests will show whether an ore carries any precious metals. Afterwards samples of the rock should be assayed to ascertain the amount of value per ton.

Gold. — Powder; roast if sulphurets are present; grind very fine and wash in pan or spoon; examine with lens; yellow particles not soluble in nitric acid. The color of pure gold is bright yellow, tinged with red. Gold may be distinguished from all other metals or alloys by the following simple traits: It is yellow, malleable, not acted upon by nitric acid.

Silver. — Pure silver is the brightest of metals, of a beautiful white color and rich luster.

Chloride of Silver. — If suspected in a pulp, harshly rub a bright and wet copper cartridge thereon. If a chloride or chloride-bromide of silver, it will whiten the copper. Graphite will thus whiten copper or gold, but can be rubbed off.

Copper. — After roasting the pulp, intimately mix and well knead with a like quantity of salt and candle grease or any other fat, and cast into the fire, when the characteristic colors — first blue, then green — will appear. This test is better made at night.

Galena. — Black zinc blende is often mistaken for galena. The two may be distinguished by the infallible sign: The powder of galena is black; that of blende, brown or yellow.

SPECIFIC GRAVITIES OF LIQUIDS LIGHTER THAN
WATER

TEMPERATURE 63⁵ F.					
DEGREES BAUMÉ	SPECIFIC GRAVITY	DEGREES BAUMÉ	SPECIFIC GRAVITY	DEGREES BAUMÉ	SPECIFIC GRAVITY
10	1.0000	27	0.8957	44	0.8111
11	0.9932	28	0.8902	45	0.8066
12	0.9865	29	0.8848	46	0.8022
13	0.9799	30	0.8795	47	0.7978
14	0.9733	31	0.8742	48	0.7935
15	0.9669	32	0.8690	49	0.7892
16	0.9605	33	0.8639	50	0.7849
17	0.9542	34	0.8588	51	0.7807
18	0.9480	35	0.8538	52	0.7766
19	0.9420	36	0.8488	53	0.7725
20	0.9359	37	0.8439	54	0.7684
21	0.9300	38	0.8391	55	0.7643
22	0.9241	39	0.8343	56	0.7604
23	0.9183	40	0.8295	57	0.7565
24	0.9125	41	0.8248	58	0.7526
25	0.9068	42	0.8202	59	0.7487
26	0.9012	43	0.8156	60	0.7449

SPECIFIC GRAVITIES OF LIQUIDS HEAVIER THAN WATER

	TEMPERATURE 63⁵ F.				
DEGREES BAUMÉ	SPECIFIC GRAVITY	DEGREES BAUMÉ	SPECIFIC GRAVITY	DEGREES BAUMÉ	SPECIFIC GRAVITY
0	1.0000	26	1.2153	52	1.5487
1	1.0068	27	1.2254	53	1.5652
2	1.0138	28	1.2357	54	1.5820
3	1.0208	29	1.2462	55	1.5993
4	1.0280	30	1.2569	56	1.6169
5	1.0353	31	1.2677	57	1.6349
6	1.0426	32	1.2788	58	1.6533
7	1.0501	33	1.2901	59	1.6721
8	1.0576	34	1.3015	60	1.6914
9	1.0653	35	1.3131	61	1.7111
10	1.0731	36	1.3250	62	1.7313
11	1.0810	37	1.3370	63	1.7520
12	1.0890	38	1.3494	64	1.7731
13	1.0972	39	1.3619	65	1.7948
14	1.1054	40	1.3746	66	1.8171
15	1.1138	41	1.3876	67	1.8398
16	1.1224	42	1.4009	68	1.8632
17	1.1310	43	1.4143	69	1.8871
18	1.1398	44	1.4281	70	1.9117
19	1.1487	45	1.4421	71	1.9370
20	1.1578	46	1.4564	72	1.9629
21	1.1670	47	1.4710	73	1.9895
22	1.1763	48	1.4860	74	2.0167
23	1.1858	49	1.5012	75	2.0449
24	1.1955	50	1.5167		
25	1.2053	51	1.5325		

COMPARISON OF THERMOMETER SCALES OF CENTIGRADE, FAHRENHEIT AND REAUMUR

C. WATER FREEZES AT 0° WATER BOILS AT 100° F. WATER FREEZES AT 32° WATER BOILS AT 212° R. WATER FREEZES AT 0° WATER BOILS AT 80°

CENTIGRADE	FAHRENHEIT	REAUMUR
$40°$	$40°$	$32°$
37^2	35	29^7
34^4	30	27^5
31^7	25	25^1
30	22	24
28^9	20	23^1
26^1	15	20^9
23^8	10	18^6
20^6	5	16^5
20	4	16
17^8	0	14^2
15^0	5	12
12^2	10	9^8
10	14	8
9^4	15	7^5
6^7	20	5^4
3^9	25	3^1
1^1	30	0^9
0	32	0
1^7	35	1^4
4^4	40	3^5
7^2	45	5^8
10	50	8
12^8	55	10^2
15^6	60	12^5
18^3	65	14^6
20	68	16
21^1	70	16^0
23^9	75	19^1
26^7	80	21^4
29^4	85	23^5
30	86	24
32^2	90	25^8
35^0	95	28
37^8	100	30^2
40	104	32
40^6	105	32^5
43^3	110	34^6
46^1	115	36^9
48^9	120	39^1

CENTIGRADE	FAHRENHEIT	REAUMUR
50 °	122 °	40 °
51^7	125	41^4
54^4	130	43^5
57^2	135	45^8
60	140	48
62^8	145	50^2
65^6	150	52^5
68^3	155	54^6
70	158	56
71^1	160	56^9
73^9	165	59^1
76^7	170	61^4
79^4	175	63^5
80	176	64
82^2	180	65^7
85^0	185	68
87^8	190	70^2
90	194	72
90^6	195	72^5
93^3	200	74^6
96^1	205	76^9
98^9	210	79^1
100	212	80
101^7	215	81^4

A variety of circumstances arise in which it becomes necessary to convert readings from one scale into those of the others, in which case the following rules are to be observed:

1. To convert Centigrade degrees into degrees of Fahrenheit, multiply by 9, divide the product by 5 and add 32.

2. To convert Fahrenheit degrees into degrees of Centigrade, subtract 32, multiply by 5 and divide by 9.

3. To convert Reaumur degrees into degrees of Fahrenheit, multiply by 9, divide by 4 and add 32.

4. To convert Fahrenheit degrees into degrees of Reaumur, subtract 32, multiply by 4 and divide by 9.

5. To convert Reaumur degrees into degrees of Centigrade, multiply by 5, and divide by 4.

6. To convert Centigrade degrees into degrees of Reaumur, multiply by 4, and divide by 5.

CYANIDES

CYANIDE OF POTASSIUM — KCN, or KCy.
CYANIDE OF SODIUM — NaCN, or NaCy.

Cyanogen is composed of Carbon and Nitrogen, both together forming a composition for which the chemical formula "Cy" is used, and which in its action is similar to one single element.

Cyan(ogen)ide of Potassium and Cyan(ogen)ide of Sodium, are combinations of Cyanogen, with the bases Potassium and Sodium, respectively, the latter only acting as carriers for the Cyanogen.

The action of the Cyanides in the extraction of gold and silver from their ores, is based on the fact that they form with these precious metals soluble double compounds.

ANTIDOTES FOR CYANIDE POISONING

All Cyanides are deadly poisons; but the aqueous solutions used in practice are so dilute that there is little or no danger from the prussic acid evolved from them if the buildings are properly ventilated.

Acids react on Cyanides, liberating prussic acid gas, which causes almost instant death when inhaled in a pure state. When diluted with air, it causes faintness, dizziness, and a depressing frontal headache.

Even very dilute solutions of Cyanide are poisonous when taken internally; and, when they come in contact with the skin, produce, in some persons, an eruption of painful red boils. In cases where the hands and arms must be brought into contact with the solutions, rubber gloves, reaching over the elbows, should be provided for the workmen. Kaffir workmen are said to suffer no inconvenience whatsoever from the contact of their skin with Cyanide solutions.

Considering the extensive use of Cyanide, the number of fatal accidents is remarkably small. Up to the present time only one fatal case has been recorded in New Zealand.

In case of accident from Cyanide poisoning, the following remedies are recommended: Put the patient into a hot bath, and apply cold water to his back and neck. In cases of internal poisoning, vomiting should be induced by emetics, or by physical means.

Freshly precipitate carbonate of iron, obtained by mixing equal quantities of Sodium Carbonate and Ferrous Sulphate, is recommended for internal use.

If the poisoning is the result of inhaling prussic acid gas, it is advisable to make the patient inhale a small quantity of chlorine gas, ammonia or ether. The chlorine gas can be quickly made and applied by sprinkling a little bleaching powder on a piece of flannel moistened with Acetic Acid, and then holding the flannel to the nostrils of the patient.

GRAM TABLE, FOR THE ASSAY OF CYANIDE SOLUTIONS

IF ½ PINT OF SOLUTION GIVES OF FINE METAL	ONE TON OF SOLUTION WILL GIVE FINE METAL		
Gram	Ozs.	Pwts.	Grs.
.0001	0	0	5.5
.0002	0	0	11
.0003	0	0	16.5
.0004	0	0	22
.0005	0	1	3.5
.0006	0	1	9
.0007	0	1	14.5
.0008	0	1	20
.0009	0	2	1.5
.0010	0	2	7
.0020	0	4	14
.0030	0	6	21
.0040	0	9	4
.0050	0	11	11
.0060	0	13	18
.0070	0	16	1
.0080	0	18	8
.0090	1	0	15
.0100	1	2	22
.0200	2	5	20
.0300	3	8	18
.0400	4	11	16
.0500	5	14	14
.0600	6	17	12
.0700	8	0	10
.0800	9	3	8
.0900	10	6	6
.1000	11	9	4
.2000	22	18	8
.3000	34	7	12
.4000	45	16	16
.5000	57	5	20
.6000	68	15	0
.7000	80	4	4
.8000	91	13	8
.9000	103	2	12
1.0000	114	11	16
2.0000	229	3	8

GRAIN TABLE, FOR THE ASSAY OF CYANIDE SOLUTIONS

IF ½ PINT OF SOLUTION GIVES OF FINE METAL	ONE TON OF SOLUTION WILL GIVE FINE METAL		
Grains.	Ozs.	Pwts.	Grs.
.001	0	0	3.5
.002	0	0	7
.003	0	0	11
.004	0	0	14.5
.005	0	0	18
.006	0	0	21.5
.007	0	1	1
.008	0	1	4.5
.009	0	1	8
.010	0	1	12
.020	0	3	0
.030	0	4	12
.040	0	6	0
.050	0	7	11
.060	0	8	23
.070	0	10	11
.080	0	11	23
.090	0	13	10
.100	0	14	22
.200	1	9	20
.300	2	4	19
.400	2	19	16
.500	3	14	14
.600	4	9	12
.700	5	4	10
.800	5	19	8
.900	6	14	6
1.000	7	9	4

PEROXIDE OF SODIUM. (DIOXIDE OF SODIUM.) Na_2O_2.

It is quite a new commercial product and is rapidly taking the place of other material as a powerful oxidizing and reducing agent. It is a stable compound, when not exposed to the air. Chemically it is that *Oxide* of the metallic elementary base *Sodium*, which contains the greatest possible quantity of *Oxygen*. Expressed in figures, Peroxide of Sodium contains 19–20% *available* Oxygen. It is a valuable material for industrial mining purposes, to supply Oxygen in large quantities quickly. The Oxygen is set free by the use of dilute Sulphuric Acid, as follows:

$$Na_2O_2 + H_2SO_4 = Na_2SO_4 + H_2O + O$$

Test with 1-10 Normal Permanganate of Potash solution.

INDEX

www.ingramcontent.com/pod-product-compliance
Lightning Source LLC
Chambersburg PA
CBHW060617200326
41521CB00007B/799